The Science of Thai Cuisine

Lists of the most popular or delicious dishes in the world always include Thai food. Sriracha sauce has gone from a dipping sauce made in a small town in Thailand to become a recognizable flavor in cuisine worldwide. With a reputation of being hot and spicy, it is not uncommon to see those who try Thai food for the first time shedding tears and sporting a red nose. Yet, the Thai national cuisine has gained a high degree of global recognition and admiration despite Thailand being a relatively small country. Is this down to sheer luck, its being an extensive work of art, or, possibly, because of scientific literacy?

The Science of Thai Cuisine: Chemical Properties and Sensory Attributes approaches the art of cooking and serving from the perspective of science and proposes the possible rationales behind Thai culinary art. With applied chemistry and sensory science, it bridges the gap between food science and culinary arts, explaining the functional properties and changes in major ingredients and techniques used in Thai cuisine.

Key Features

- Discusses the chemistry of ingredients and techniques in Thai cuisine with possibilities of application and innovation
- Presents scientific research combined with the arts and history of Thai food
- Provides scientific evidence linking Thai food with the sensory perception and the joy of eating
- Contains vibrant color photographs of Thai cuisine

While there are numerous cookbooks that feature Thai cuisine, none are as dedicated as this to explaining the science behind the ingredients, cooking methods, and sensory aspects. This book will be beneficial to professionals in the food industry, appealing to chefs, food scientists, and sensory analysis experts, as well as anyone who has an interest in Thai culture.

The Science of Thai Cuisine
Chemical Properties and Sensory Attributes

Edited by
Valeeratana K. Sinsawasdi,
Nithiya Rattanapanone, and
Holger Y. Toschka

CRC Press
Taylor & Francis Group
Boca Raton London New York

CRC Press is an imprint of the
Taylor & Francis Group, an **informa** business

First edition published 2023
by CRC Press
6000 Broken Sound Parkway NW, Suite 300, Boca Raton, FL 33487-2742

and by CRC Press
4 Park Square, Milton Park, Abingdon, Oxon, OX14 4RN

CRC Press is an imprint of Taylor & Francis Group, LLC

Library of Congress Cataloging-in-Publication Data

Names: Sinsawasdi, Valeeratana K., author. | Rattanapanone, Nithiya, author. | Toschka, Holger Y., author.
Title: The science of Thai cuisine : chemical properties and sensory attributes / Valeeratana K. Sinsawasdi, Nithiya Rattanapanone, Holger Y. Toschka.
Description: First edition. | Boca Raton : CRC Press, 2022. | Includes bibliographical references and index.
Identifiers: LCCN 2021061477 (print) | LCCN 2021061478 (ebook) | ISBN 9781032023281 (hardback) | ISBN 9780367763350 (paperback) | ISBN 9781003182924 (ebook)
Subjects: LCSH: Food--Sensory evaluation. | Cooking, Thai. | Food habits--Thailand.
Classification: LCC TX546 .S56 2022 (print) | LCC TX546 (ebook) | DDC 664/.072--dc23/eng/20220112
LC record available at https://lccn.loc.gov/2021061477
LC ebook record available at https://lccn.loc.gov/2021061478

ISBN: 978-1-032-02328-1 (hbk)
ISBN: 978-0-367-76335-0 (pbk)
ISBN: 978-1-003-18292-4 (ebk)

DOI: 10.1201/9781003182924

Typeset in Caslon
by KnowledgeWorks Global Ltd.

Contents

V

Preface

The intersection of Thai Kitchen Wisdom and Science—The flow of this book.

Scientific principles behind local and ancient pearls of wisdom associated with several Thai dishes may be interesting to many. *The Science of Thai Cuisine: Chemical Properties and Sensory Attributes* begins with Part I, which provides the identity of the cuisine in Chapter 1, followed by the historical perspective of Thai food in Chapter 2. The chapters discuss the evolution and development to understand and distinguish Thai food cultures from others.

Part II, Multi-Sensory Properties of Thai Foods and Their Sources, is arranged as a sequence of a transformation, from raw materials to the consumers' consumption experience. Chapter 3 lays out the sensory attributes provided by food components and interactions. The details of compositions, properties, functions, and the characteristics of essential Thai food ingredients, such as rice and fish sauce, are in Chapter 4. To cook into a dish, Chapter 5 explains the changes that have taken place in various cooking methods, such as boiling and curry making. Since Thai meals are composed of all food items served all at the same time, the total experience of eating Thai food is as important as the flavor arising from any individual dish.

Then, Part III is the integration of Culinary Arts and Food Science. Starting with Chapter 6, possible explanations of how Thai cuisine earned its reputation for deliciousness are explored with applicable theories and scientific evidence. Next, in Chapter 7, some famous Thai dishes' ingredients and cooking methods are demonstrated in various *gub-khao* recipes. Finally, the book ends with Chapter 8, with the authors' perspectives on Thai cuisine.

About the Editors

Valeeratana Kalani Sinsawasdi grew up in Thailand. Her interest in food started at a very young age as she liked to observe her "khun-ya," making the meal plan every day. The cooking was labor intensive, yet it was fascinating. She was impressed by how herbs and spices would become invisible in the finished dish but easily detectable through taste and smell. It seems as if the preparation of curry paste was the making of a magic potion.

With her other interest in science, Valeeratana has committed herself to the field of Food Science. First, she received her bachelor's degree in Food Science from Chiang Mai University, Thailand. Then, a master's degree from the University of Hawaii and a PhD degree from the University of Florida, USA, all in Food Science.

Valeeratana has extensive experience in the food industry. In the USA, with Dean Foods, in Asia and Thailand with Unilever, and as a consultant to various companies. In the academic sphere, she has taught several Food Science courses at Mahidol University, Thailand.

Holger Y. Toschka studied Biochemistry and Molecular Biology at the Institute for Biochemistry at Free University in Berlin, focusing on gene structure and expressing of ribosomal genes.

He received his PhD in that research field in 1988 and joined Unilever after a short Post-Doc stint in early 1990 as scientist manager for Gene Technology and Fermentation in the Netherlands. Since then, he has worked at various locations in various roles in R&D and supply chain for the Food and Refreshment Divison of the company. Since the beginning of 2018, he has been the Foods R&D director for South East Asia and is based in Bangkok. It's the second time he has lived in Thailand, which gave him seven years to learn about Thai food and cooking. Holger worked until 2022 as R&D director in Bangkok. Holger is German, married to Ute, and with four fully grown children.

Nithiya Rattanapanone studied in Thailand and received her BSc and MSc from Chiang Mai University and Mahidol University. She then earned her PhD in Food Science from the University of Nottingham, England, in 1977 for her pioneering research on mRNA changes during the ripening of tomatoes. She extended her interest to many topics of Food Science and published more than 70 peer-reviewed articles during her time teaching at the Department of Food Science, Chiang Mai University. She also authored many books and textbooks in Food Chemistry, Post-Harvest Technology, and Human Nutrition, notably for college-level courses in Thailand. In addition, Nithiya has served as visiting scientist at many international institutes, for example, the USDA Agricultural Research Service, Beltsville, Maryland, the Citrus and Other Products Research, Florida, and the FDA National Center for Toxicological Research, Arkansas, in the USA; CIRAD-FLHOR, Montpellier, in France; and at the Faculty of Food Science and Technology, University of Reading, in the UK.

In 2002, she became the first lecturer in Thailand to receive the academic rank of Professor in Food Science and Technology. Then, she was honored as Professor Emeritus for Chiang Mai University after she retired in 2011. To date, she actively contributes to academics and the Thai government agency.

List of Contributors

Nate-tra Dhevabanchachai, D.HTM
Human Resources Development
    and Service Excellence,
    Asia Pacific
Bangkok, Thailand

Nithiya Rattanapanone, PhD
Chiang Mai University
Chiang Mai, Thailand

Narong Sinsawasdi, PhD
Chiang Mai University
Bangkok, Thailand

Valeeratana Kalani Sinsawasdi, PhD
Vingenious, LLC
Los Angeles, CA, USA
(formerly faculty member,
    Mahidol University, Thailand)

Holger York Toschka, PhD
Unilever
Bangkok, Thailand

PART I
THE THAI FOOD CULTURE

1

THAI CUISINE IDENTITY

VALEERATANA K. SINSAWASDI AND NATE-TRA DHEVABANCHACHAI

Contents

1.1 Introduction

Thailand is a country in Southeast Asia having about 70 million people with about 5 hundred thousand square kilometers of land (slightly larger than California, the United States). Though being relatively small, the country has several terrains, consisting of mountains in the north and the west, plateaus in the northeast, sedimentary plains in the central region, and coastal plains and islands in the south. The differences in climate, soil, and water resources contribute to a wide array of edible plants and animals.

Most of the people consider themselves Thai, although each region has its own identity, especially when it comes to food culture. The country was modernized, especially during the nineteenth century, to embrace Western knowledge and culture. However, Thailand or Siam was able to retain its independence even during Western imperialism (Royal Thai Embassy 2021; Tourism Authority of Thailand 2021). Thus, Thai culture has not been directly influenced by any other nations (Sinsawasdi 1998). The excerpt from a 1930 film, produced by Associated British Film Distributors Ltd. titled "I am from Siam," is still accurate to describe Thailand today: "Here we have a strange mixture of East and West, ancient colorful ceremonies side by side

DOI: 10.1201/9781003182924-2

with the most modern mechanical improvements." "Magnificent and spectacular ceremonies. Bangkok, the Venice of the East. The jewels of Asia" (Garden et al. 1930).

1.2 Thai Cuisine and the World: Perception and Identity

Thai food is a significant international cuisine. It is very popular among foreign tourists visiting Thailand as well as diners visiting Thai restaurants abroad. Traditional dishes like *tom-yum-koong, som tum,* and *massaman curry* have been recognized, and frequently appear on the list of the most delicious foods in the world. In addition to culinary dishes, Thai flavors, such as Sriracha hot sauce, have recently become a household name and been selected to flavor several processed foods and snacks worldwide.

Thai cuisine has been declared as the most flavorful food in the world (Nachay 2019). A simple Internet search for the phrase will result in several web pages ranking the most delicious food and Thai cuisine remains often among the top choices. In 2017, CNN, one of the largest international news media, conducted a poll to vote for the world's best foods. The campaign was well responded to, with more than 35,000 voters from the whole world giving their opinions on what they thought was the best dish. Although, out of the 50 dishes on the final list, Thai food was not ranked at the top, it astonishingly occupied the most places on the list with seven dishes. The dishes, ranked according to the highest number of votes, were ***tom-yum-koong*** (spicy soup with shrimp, ต้มยำกุ้ง), **pad thai** (stir-fried rice noodles, ผัดไทย), ***som-tum*** (spicy green papaya salad, ส้มตำ), **massaman curry** (*gaeng-mut-sa-mun,* แกงมัสมั่น), **green curry** (*gaeng-keow-wan,* แกงเขียวหวาน), **fried rice** (*khao-pud,* ข้าวผัด), and ***moo-nam-tog*** (spicy pork salad, หมูน้ำตก) (Cheung 2017). However, when the list was updated in 2019 and 2021, the massaman curry ranked first. The other two dishes on the list were ***som-tum*** (spicy green papaya salad) and ***tom-yum-koong*** (spicy soup with shrimp) (CNN Travel Staff 2019, 2021).

As per another CNN feature on this topic, Thailand was listed among the top ten countries with the best food (Li 2019). It is not only famous for providing fine dining in a restaurant, but also for providing comparatively cheap foods by mobile vendors. Bangkok was listed as the best city in the world for the availability of street food.

Recommended dishes from the street of Bangkok are, for example, **pad thai** (stir-fried rice noodle, ผัดไทย), *hoy-tod* (oyster crispy omelet, หอยทอด), and **chicken noodle soup** (Shea 2018). Bangkok was ranked number 3 on the list of the top cities for dining in tourists spending on foods, indicating Thailand's food tourism trend (TAT News Room 2019).

As for Thai food offered in Thai restaurants overseas, one study indicated that American consumers accepted Thai food as healthy, aromatic, and of good quality (Jang, Ha, and Silkes 2009). There are as many as 12,000 Thai restaurants in foreign countries, according to the Department of International Trade Promotion, Ministry of Commerce, Thailand (Prachachat Business 2021). In a survey of frequently ordered dishes from Thai restaurants overseas, the top ten dishes (ranked from the most frequent to the least) were ***tom-yum-koong*** (spicy soup with shrimp, ต้มยำกุ้ง), **green curry** (*gaeng-keow-wan*, แกงเขียวหวาน), **pad thai** (stir-fried rice noodle, ผัดไทย), **spicy stir-fried minced meat with holy basil** (*pud-ka-prao*, ผัดกะเพรา), **roasted duck red curry** (*gaeng-ped-pade-yang*, แกงเผ็ดเป็ดย่าง), ***tom-kha-gai*** (spicy galangal soup with chicken, ต้มข่าไก่), **spicy beef salad** (*yum-nua*, ยำเนื้อ), **pork satay** (*moo-sa-tay*, หมูสะเต๊ะ), **stir-fried chicken with cashew nuts** (*gai-pud-med-ma-muang-him-ma-pan*, ไก่ผัดเม็ดมะม่วงหิมพานต์), and *pa-nang* (red curry with coriander seeds and peanuts, พะแนง). The number of participating Thai restaurants in this survey was more than 1,500, including restaurants from America, Europe, Asia, and Australia (Chavasit et al. 2003).

A survey in the United States by Mintel on international food preferences reflects the transition toward Thai food among the younger generation (millennials) compared with baby boomers. One of the key reasons for which tourists visit Thailand is its culinary arts, widely considered a signature experience (Suntikul 2019; Walter 2017). Popular foods among tourists and expatriates are ***tom-yum-koong*** (spicy shrimp soup, ต้มยำกุ้ง), **spicy stir-fried holy basil with minced meat** (*pud-ka-prao*, ผัดกะเพรา), *som-tum* (spicy papaya salad, ส้มตำ), and **grilled chicken** (*gai-yang*, ไก่ย่าง). These dishes are simple and widely available even on streets. For many tourists, street food is perceived as a Thai culture presentation (Agoda 2019; Beijers 2021; Henderson 2019). Additional experience shared by multi-nationality expatriates who had lived in Thailand is listed in Appendix A. Most expatriates started to like Thai food from their maiden experience and sustained their preferences even after leaving Thailand.

Thai flavors in supermarket food products and restaurant dishes such as Sriracha and sweet Thai chili sauce have been on the list of trending flavors (Nachay 2017; Sloan 2019). While Thailand's population makes up less than 1% of the world population, it is a phenomenon how Thai cooking and eating have gained fame and popularity globally.

Padoongpatt (2017) studied the origin of Thai restaurants in Los Angeles and observed that the Thai cuisine "boom" started in the 1970s. By referring to many sources and literature, Padoongpatt pointed out that the location of Thai restaurants played an important role for making Thai cuisine so popular. With many famous Thai restaurants close to big-name studios such as the 20th Century Fox, Hollywood stars and celebrities like Madonna and Harrison Ford were acclaimed regulars (Sodsook 1995). Then in the 1980s, when weight-loss food was in trend, Thai foods gained yet another identity advantage. In addition to the deliciousness suited even for celebrities, Thai cuisine was recognized as healthier and lighter than other Asian foods (Brennan 1981; Van Esterik 1992).

In Thai restaurants in the UK, menu item descriptions containing nationalism elements (e.g. Thai steamed sticky rice or use of local words) appeared the most appealing to customers. Sensory cues and dish characteristics were also influential, with dominant keywords identified as curry, chili, and spicy (Low, 2021). As restaurant branding and health implications were the least recognized for appealing perception, the research highlighted the importance of authentic Thai foods characterized by spiciness. Interestingly, especially on the aspect of nationalism, a dish named pad thai is no longer a transliteration. The word "pad thai" has now been officially included in Oxford Dictionary as an English word (www.oxfordlearnersdictionaries.com).

1.3 The National Aspect of Culinary Culture: Thai Foods According to Thai People

A great Thai meal is a pure and honest reflection of the countries' long and unique history, geographical location, climate, and biological diversity. These mixtures of factors help one to build the unique and very distinctive Thai food tradition that nowadays is easily distinguishable from other national cuisines.

Although it is generally known as Thai cuisine, it can be categorized further into several regional ones. The central Thai cuisine is the cuisine of the flat and wet central rice growing plains. Its typical meal is composed of spicy soup, clear soup, deep-fry or stir-fry, and chili paste. The dishes are milder and sweeter than those from other regions.

Apart from that, it is the Isan or northeastern Thai cuisine that more reflects the essence of the arid Khorat Plateau and its food sources. These are similar to the culture of Laos, enriched by the Khmer cuisine. *Pla-ra* (fermented freshwater fish, ปลาร้า) is often quoted as a lead representation, along with the hotness of chili.

The Northern Thai cuisine finds its roots in the former Lanna kingdom and mirrors the landscape of the mountains of the Thai Highland and its cool valleys. Cooking oil is an essential ingredient as fried garlic is used in several dishes. However, coconut milk is not used in this cuisine, possibly because coconut trees are not grown in the high mountainous region.

Thai cuisine of the south, which frequently includes coconut milk and turmeric, is influenced by Malay, Indian, Indonesian, and Cantonese culture. Hainanese elements can also be experienced. The cuisine can be referred to as *Ya-ya* or Peranakan and is famous for its hotness and saltiness (Kongpan 2018; Svasti 2010; The Department of Cultural Promotion-Ministry of Culture 2021).

The cultural aspect of Thai cooking and eating patterns is honored as a cultural feature. The uniqueness of how Thai people cook and eat their foods has been recognized and certified on the list of intangible cultural heritage. It is an attempt led by the Thailand Department of Cultural Promotion at the Ministry of Culture to present Thai food on UNESCO's list of certified intangible cultural heritage (The Department of Cultural Promotion-Ministry of Culture 2021; UNESCO 2019). The certification has been updated every year since it started in 2012. More details can be found on the department's website (http://ich.culture.go.th).

Thai culinary delicacies that have been on the cultural heritage list are, for example, ***tom-yum-koong*** (ต้มยำกุ้ง), **pad thai** (ผัดไทย), **red curry** (gaeng-ped, แกงเผ็ด), **green curry** (*gaeng-keow-wan*, แกงเขียวหวาน), **chili** dip (*nam-prig*, น้ำพริก), and **spicy green papaya salad** (*som-tum*, ส้มตำ). Besides individual dishes, Thai custom of meal ensemble and table setting called ***sum-rub*** (สำรับ) has also been certified as cultural heritage.

Thai people eat rice as a primary source of energy, traditionally in every meal. The main savory dish to be eaten with rice is called ***gub-khao*** (กับข้าว), which has direct translation as (to be eaten) "with rice." These *gub-khao* dishes are compatible with main courses in a Western-style meal. The term reflected a perception of rice as the pivotal item, and everything else has an inferior role of rice accompaniment. This concept is the opposite of the Western-style meal, which regards cooked rice as a side dish. Thai people continue to call these savory dishes *gub-khao* (culinary repertoire to be eaten with rice) even though, probably started in 1940 with a government campaign on national nutrition status, the proportion of rice eaten in a meal has been decreasing (Puaksom 2017).

The meal setting with *gub-khao* varieties takes a relatively long time to prepare and eat. Therefore, a one-plate dish is preferred, especially for lunchtime. Pad thai is an example of a one-plate dish. Other *gub-khao* such as massaman or green curry can be served as a quick meal by placing the food on the same plate of rice, hence the name ***khao-rad-gaeng*** (curry-topped rice, ข้าวราดแกง). A survey by university students showed that the most popular dishes for lunch are all one-plate meals. These dishes were **chicken rice** (*khao-mun-gai*, ข้าวมันไก่), **noodles with pork** (*guay-teow-moo*, ก๋วยเตี๋ยวหมู), **spicy stir-fried basil with crispy fried egg** (*khao-ka-prao-kai-dow*, ข้าวกระเพราไข่ดาว), **rice with a savory omelet** (khao-kai-jeow, ข้าวไข่เจียว), and **fried rice** (*khao-pad*, ข้าวผัด) (Rapaphan et al. 2014). Except for the spicy stir-fry, all the food has a mild taste with no or low use of herbs and spices.

During dinner time, more complicated food is preferred. Soupy *gub-khao* is called ***gaeng*** (แกง) or ***tom*** (ต้ม). The soupy recipes can be classified further by ingredients and flavor intensity, from the most intense flavor such as *gaeng* with coconut milk and herbs and spices (e.g., green curry) to the broth-based soup like ***gaeng-jued*** (e.g., cabbage soup with minced pork). Food groups with more concentrated herbs and spices, especially chili, are **chili dip** (*nam-prig*, น้ำพริก) and chili dip with coconut milk (*lhon*, หลน). Many Thai foods require simple tossing or mixing, mostly with spicy seasoning, such as the **green papaya salad** (*som-tum*, ส้มตำ). Stir-frying is another popular cooking technique. The stir-fried dishes can be spicy such as **spicy stir-fried holy basil with minced meat** (*pud-ka-prao*, ผัดกะเพรา) and can have mild taste such as those in varieties of **stir-fried vegetables** (*pud-pug*, ผัดผัก).

The **Thai-style meal setting** (*sum-rub*) typically comprises one chili dip with vegetables, one meat dish such as fried fish, one spicy, soupy dish, one clear soup, and either stir-fried dish (with or without spiciness) or spicy salad. All savory dishes are served in all-at-once manner together with rice (Kongpan 2018; The Department of Cultural Promotion-Ministry of Culture 2021). The number and type of dishes are flexible and depend on the number and age of diners, and occasion of eating. See also Chapter 6 for further culinary details of the *gub-khao* and *sum-rub*.

In a survey on frequently consumed ***gub-khao*** (savory dishes) among Thai people living in Thailand, the top five dishes were **stir-fried veg-etable** (*pud-pug*, ผัดผัก), **ivy gourd soup** (*gaeng-jued-tum-lueng*, แกงจืดตำลึง), **shrimp paste chili dip** (*nam-prig-ga-pi*, น้ำพริกกะปิ), **spicy stir-fried holy basil with chicken** (*pud-ka-prao-gai*, ผัดกะเพราไก่), and **green chicken curry** (*gaeng-keow-wan-gai*, แกงเขียวหวานไก่) (Maruekin 2002). These dishes represent typical *sum-rub* dinner settings in Thai households. The shrimp-paste chili dip is always accompanied by a lot of vegetables and fried fish. Milder vegetable soup and vegetable stir-fries provide micronutrients and fiber and help to refresh the palate after spicy foods.

Although spiciness has become an essential characteristic of Thai foods, anyone who despises hotness or piquancy can easily enjoy Thai food throughout the whole meal without any chili or spices on their plate. A comparison of the top five dishes from different contexts is shown in Table 1.1. All the dishes listed in the table, except the

Table 1.1 Five Most Popular Thai Dishes

RANK	THAI RESTAURANTS OVERSEAS[a]	WORLDWIDE VOTES[b]	THAI PEOPLE IN THAILAND[c]
1	*Tom-yum-koong*	*Tom-yum-koong*	Stir-fried vegetables
2	Green curry	Pad thai	Ivy gourd soup
3	Pad thai	Spicy green papaya salad	Shrimp-paste chili dip
4	Spicy stir-fried holy basil with meat	Massaman curry	Spicy stir-fried holy basil and chicken
5	Roasted duck red curry	Green curry	Chicken green curry

[a] Chavasit et al. (2003).
[b] Cheung (2017).
[c] Maruekin (2002).

vegetable soup and stir-fried, have been certified with UNESCO as intangible cultural heritage of Thailand.

1.4 Conclusion

Though the area is small and the population is not high, the supportive climate and geographic location give Thailand many advantages in developing its national cuisine. As a result, there are wide varieties of ingredients and combinations of regional foods. Globally, Thai food has been well recognized and continued to gain a reputation for its deliciousness, especially the iconic massaman curry, *som-tum*, and pad thai. Although spicy dishes with chili seem to symbolize Thai food, Thai food culture is more intricate. In a typical meal, Thai people ensemble various dishes (*gub-khao*) into a *sum-rub*-style table setting for a balance of flavors and textures in a meal.

Acknowledgement

We would like to express our sincere gratitude to Ms. Rosanaporn Viravan, former lecturer at the Faculty of Arts, Chulalongkorn University, for her invaluable guidance in the transcription of Thai names to English.

References

Agoda. 2019. "Bangkok Food: Can't Miss Thai Dishes & Best Places to Eat Street Food." https://www.agoda.com/travel-guides/thailand/bangkok/bangkok-food-cant-miss-thai-dishes-best-places-to-eat-street-food?cid=1844104.

Beijers, Joost. 2021. "14 Best Thai Food For Foreigners 2021." https://toptravelfoods.com/article/thai-food-for-foreigner.

Brennan, Jennifer. 1981. *Original Thai Cookbook*. New York, USA: Richard Marek.

Chavasit, Visith, Vongsvat Kosulwat, Somsri Charoenkiatkul, and Somkiat Kosulwut. 2003. "Exporting Thai Restaurant Business." In *Opportunities and Potentials of Developing Thai Food to the World Market* (การวิเคราะห์โอกาสและศักยภาพของการพัฒนาอาหารไทยสู่ตลาดโลก), 41–43. Bangkok, Thailand: Thailand Science Research and Innovation (TSRI).

Cheung, Tim. 2017. "Your Pick: World's 50 Best Foods." CNN, accessed January 8. https://edition.cnn.com/travel/article/world-best-foods-readers-choice/index.html.

CNN Travel Staff. 2019. "Culinary Journeys, The World's 50 Best Foods." CNN, accessed January 8, Last Modified https://edition.cnn.com/travel/article/world-best-food-dishes/index.html.

CNN Travel Staff. 2021. "The World's 50 Best Foods." https://www.cnn.com/travel/article/world-best-food-dishes/index.html.

Garden, D.S., Russell Shilds, M. J. Weisfeldt. 1930. *I am from Siam*. U.K.: Associated British Film Distributors Ltd.

Henderson, Joan C. 2019. "Street food and tourism: A Southeast Asian perspective." In *Food Tourism in Asia*, 45–57. Singapore: Springer.

Jang, SooCheong Shawn, Aejin Ha, and Carol A Silkes. 2009. "Perceived attributes of Asian foods: From the perspective of the American customers." *International Journal of Hospitality Management* 28 (1):63–70.

Kongpan, Srisamorn. 2018. *Intangible Cultural Heritage Foods of Thailand* (อาหารขึ้นทะเบียน มรดกทางภูมิปัญญาของชาติ) Bangkok, Thailand: S.S.S.S. (บริษัท ส.ส.ส. จำกัด).

Li, Zoe. 2019. "Which Country has the Best Food?." CNN, accessed January 8. https://edition.cnn.com/travel/article/world-best-food-cultures/index.html.

Low, P. (2021). "Descriptive language of'Thai SELECT premium'restaurant menus: Appealing perception and collocations." *Journal of Language and Linguistic Studies*, 17 (4), 1669–1683.

Maruekin, Pattaya. 2002. *Development of Reference Recipes for Commonly Consumed Thai Side Dishes and their Nutritive Values*. Bangkok, Thailand: Mahidol University.

Nachay, Karen. "Travel the Globe One Ingredient at a Time." FOOD TECHNOLOGY 71, no. 9 (2017): 55.

Nachay, Karen. "Fresh spins on global cuisines." FOOD TECHNOLOGY 73, no. 7 (2019): 48–55.

Padoongpatt, Mark. 2017. "3. Too Hot to Handle? Restaurants and Thai American Identity." In *Flavors of Empire*, 85–117. California, USA: University of California Press.

Prachachat Business. 2021. "Thailand Kitchen of the World and THAI SELECT." Matichon. https://www.prachachat.net/economy/news-632561.

Puaksom, Davisakd. 2017. "National Food Policy 1940 (โครงการส่งเสริมอาหารของชาติยุคจอมพล ป. 2481 'บันไดขั้นแรกของการสร้างชาติ")." Hfocus. https://www.hfocus.org/content/2017/07/14292.

Rapaphan, Ananda, Nipavada Chomchome, Karunee Kwanbunjan, and Valeeratana Kalani Sinsawasdi. 2014. "Food Choices and Body Mass Index of Mahidol University Students, a Nutrition survey on Salaya campus." The 1st National Conference on Food and Nutrition for Health "Fruits and Vegetables for Nutrition Security," Bangkok, Thailand, 16–17 January 2014.

Royal Thai Embassy. 2021. "Thailand in Brief." https://thaiembdc.org/about-thailand/thailand-in-brief/.

Shea, Griffin. 2018. "Best 23 Cities for Street food from Miami to Tokyo." https://edition.cnn.com/travel/article/best-cities-street-food/index.html.

Sinsawasdi, Narong. 1998. *Psychology of Thai Politics* (การเมืองไทย: การวิเคราะห์เชิง จิตวิทยา). Bangkok Thailand: Oriental Scholar.

Sloan, A Elizabeth. 2019. "A New Wave of Asian Cuisine." Food Technology Magazine.

Sodsook, Victor. 1995. *I Love Thai Food*. Los Angeles, California: Spice Market Studio.

Suntikul, Wantanee. 2019. "Gastrodiplomacy in tourism." *Current Issues in Tourism* 22 (9):1076–1094. doi: 10.1080/13683500.2017.1363723.

Svasti, Sirichalerm. 2010. *The Principles of Thai Cookery*. Nonthaburi, Thailand: McDang dot Com Company, Limited.

TAT News Room. 2019. "Bangkok Ranked in Top Ten List of Global Cities for Dining and Shopping in MasterCard Index 2018." https://www.tat-news.org/2019/01/bangkok-ranked-in-top-ten-list-of-global-cities-for-dining-and-shopping-in-mastercard-index-2018/.

The Department of Cultural Promotion-Ministry of Culture. 2021. *Intangible Cultural Heritage*. Bangkok, Thailand: Food and Nutrition.

Tourism Authority of Thailand. 2021. "History & Geography & Geology." https://www.tourismthailand.org/Articles/plan-your-trip-history-and-geography-geology.

UNESCO. 2019. "Intangible Cultural Heritage." https://ich.unesco.org/en.

Van Esterik, Penny. 1992. "From Marco Polo to McDonald's: Thai cuisine in transition." *Food and Foodways* 5 (2):177–193. doi: 10.1080/07409710.1992.9961999.

Walter, Pierre. 2017. "Culinary tourism as living history: Staging, tourist performance and perceptions of authenticity in a Thai cooking school." *Journal of Heritage Tourism* 12 (4):365–379. doi: 10.1080/1743873X.2016.1207651.

2

HISTORICAL PERSPECTIVE OF THAI CUISINE

VALEERATANA K. SINSAWASDI AND NARONG SINSAWASDI

Contents

2.1 Introduction

Present-day Thailand can be traced back to the establishment of the Kingdom of Sukhothai in the year 1238. The center of this kingdom, Sukhothai, was in the northern part of the country. Then in the year 1350, another kingdom, Ayutthaya, was established. The city of Ayutthaya was in the country's central region, about 300 km south of Sukhothai. Ayutthaya period lasted for 427 years. Then in 1782, Bangkok was established as the new capital city by King Rama I of the Chakri Dynasty. Bangkok was about 70 km south of Ayutthaya. Successive kings of this dynasty would rule Thailand to the present time (Dhiravegin 2002).

In regards to the diet, people in this area have eaten mainly rice and fish for over 3000 years. There were plenty of rice paddies with enough water, and fish was abundant in any water resources. The land was described as "in the paddies, there is rice-in the water there is fish"

DOI: 10.1201/9781003182924-3

(ในน้ำมีปลาในนามีข้าว) since the Sukhothai period (13th century). In addition, there was evidence for soupy dishes with meat called gaeng (แกง) (Kanchanakunjara et al. 2015). Desserts existed with ingredients such as rice and honey. There were plenty of larger land animals, including cows, which could easily be slaughtered for consumption. However, probably influenced by Buddhism, people did not eat land animals to that extent (Castillo et al. 2016; Teanpewroj 2017; Wongthes 2017).

The rice and fish have been synonyms for "food". Examples are in phrases like, *gin-khao-gin-plaa* (eat rice – eat fish, กินข้าวกินปลา), *gub-khao-gub-pla* (rice and savory dishes, กับข้าวกับปลา), *kao-pla-ar-harn* (rice, fish, food, ข้าวปลาอาหาร). Salt is also available, especially from the northeastern part of the country. Archaeological evidence confirmed both the salt mining and fermentation of fish with salt (Weber et al. 2010; Yankowski, Kerdsap, and Chang 2015).

Over the following millennia, rice, salt, and fish continue to be vital food for Thai people. Noteworthy evidence is in the Bowring Treaty that Siam signed with the British in 1855 as part of an attempt to remain independent during colonialism. Among the many disadvantages of trade deals the country was pressured to accept, King Mongkut (Rama VI) firmly stated an exception that no rice, fish, and salt were to be exported if they were not enough for local consumption (Bruce 1969).

The best earliest known description of Thai cuisine is in the poem *Boat Song Admiring the* Savory and the Sweets (*karp-he-chom-kruang-kow-wan*, กาพย์เห่ชมเครื่องคาวหวาน) composed by King Rama II around the late 18th century (Wongcha-um 2009). Although this poetry work did not include recipes and techniques, its list of Thai food items and their characteristics have been used as a significant reference for Thai food history.

The legendary Thai food cookbook was titled *Mae-Krua-Hua-Pa* (The Lady Chef, แม่ครัวหัวป่าก์), written by Thanphuying Plian Phassakorawong (Bunnag). It is regarded as the first comprehensive Thai cookbook and included other general topics of home economics as well. *Mae-Krua-Hua-Pa* cookbook was first published as journal articles in 1888, then later as a book and is still widely available commercially in bookstores to this day (Kasempolkoon, 2017).

2.2 Basis of Thai Food Heritage; Fusion of International Food Tradition

The area of Thailand had long been a part of the Maritime Silk Roads. Besides, waterways running to the Gulf of Siam have also facilitated international trades. Thus, for centuries, Thai people have traded with merchants and visitors from many regions, such as the Mediterranean area, India, China, and the Middle East (Castillo, Bellina, and Fuller 2016; Srisophon 2018; UNESCO 2021). Visitors from several Asian and European countries came for trade and introduced several overseas food items to Thailand. The wine was imported from Spain, spices from India, desserts from Portuguese and Chinese cuisines (Nitivorakarn 2014). Though the pungency heat of chili pepper seems to symbolize Thai cuisine, the plant actually originated in South America. Portuguese traders introduced chilies to the Thai people around the 16th century (Sukphisit 2019). Papaya was also brought from Central America during the Spanish exploration and eventually to Southeast Asia and Thailand in the 18th century (Wongthes 2017; Yeh et al. 2007). Thailand has a similar climate as South America and Indonesia regarding rainfalls and temperature (Kim and Chung 2016). Thus, many plant-based ingredients that had to be imported from these regions eventually became naturalized plants.

Interestingly, most Thai people's diet and food records in the Ayutthaya period were compiled by foreigners visiting the Kingdom. Therefore, it has been hypothesized that food and cuisine were rarely recorded in Thai literature because the food was always plentiful, and there were abundant rice, fish, and produce varieties. However, this scenario changed during the long famine caused by the war that marked the end of the Kingdom. During the time, people were starving, and food gained much more interest and priority. Thai people started developing, recording, and incorporating recipes and menus into literature and documentaries, during the early Chakri Dynasty of the Rattanakosin Kingdom (Sujachaya 2017).

The basis of Thai culinary heritage has been built around the fusion of several food traditions over many centuries. The incorporation of several foods and cultures from many regions with a mind open to embrace new ideas fused into creation of foods that later became the identity of Thai cuisine. Ayutthaya era is recognized as the golden era of Thai food development. With visitors from so many nations and races, the Siamese or Thai people classified foreigners into three main groups: *Fa-rang*, *Khaek*, and *Chin*.

The fusion basis of Thai cuisine was evident in the first Thai cookbook, *Mae-Krua-Hua-Pa*, which has also been honored as a reference for authentic Thai cuisine recipes. The names of dishes in this book reflect how Thai cuisine has been ingeniously developed with the influences of many other ethnic groups. The nations specified in the dish names include Vietnamese, Cambodian, Laotian, Chinese, Indian, Arab, Italian, English, and American. Another notable feature in this cookbook is that there are over 50 recipes with the indication of spiciness in the name, such as curry (*gaeng*, แกง) and spicy (*ped*, เผ็ด). These bold and spicy flavor recipes are always eaten with rice and assorted vegetables (Pewporchai 2017; Phassakorawong 1910).

2.3 European Influence

Fa-rang (ฝรั่ง) refers to Caucasians from Europe, with the Portuguese as the most prominent in the history of Siam. A lady of Portuguese descent named Marie Pihna de Guimar (ท้าวทองกีบม้า), who worked for the Ayutthaya royal court in the 17th century, has been credited as the person who introduced milk, egg, butter, and the technique of baking in a brick oven to Siam. The recipes, such as *foi-thong* (similar to fios de ovos or Portuguese egg thread, ฝอยทอง) and *thong-yod* (sweeten egg yolk drops, ทองหยอด), were developed with some adaptations to local ingredients and preferences till they became well-respected national Thai desserts (Suppakul et al. 2016).

Volume 4 of the *Mae-Krua-Hua-Pa* cookbook, widely credited as the first Thai cookbook, contains three macaroni and pasta recipes. There was macaroni in clear soup, thick soup, and white sauce. The author mentioned the origin of pasta as from Italy. The soup was Thai style with garlic, pepper, and coriander root mixed with meat and seasoned with fish sauce. In another recipe, the white sauce ingredients included cream, milk, and cheese. Nowadays, Thai people, especially children, consume more dairy products, especially milk and yogurt. Still to this day, the Western-style food from *Fa-rang* (Caucasians), especially milk, is believed to help kids grow taller (Sinsawasdi et al. 2020).

In volume 5, three recipes have the Western ingredient used with traditional Thai dishes, such as canned turkey, canned ham, and anchovies. The first one is spicy turkey and ham dip (*plah-gai-nguang-ham*, พล่าไก่งวงแฮม), second spicy turkey and ham in coconut milk (*gai-nguang-ham-lon*, ไก่งวงแฮมหลน), and third spicy anchovies dip (*plah-plar-anchovies*, พล่าปลาแอนโชวี) (Phassakorawong 1910). The spicy anchovy dip is the most aromatic. It contains many herbs and spices, i.e., lemongrass, ginger, shallot, kaffir lime juice, chili peppers, coriander leaves, kaffir lime leaves, and palm sugar. The cookbook recommended the spicy anchovy dip to be eaten with a lot of vegetables and grilled fish or charbroiled shrimps.

2.4 Islamic Influence

The nomenclature *Khaek* (แขก) was associated with those foreigners under Islamic influence, such as Persians, Middle Easterners, South Asians, Indo-Malays, and Asian Muslims. Massaman, a curry with several spices (cardamom, cumin, nutmeg, etc.), has been traced to Persian merchants since the 16th century. One of the Persian visitors continued to live in Siam and established the Bunnag family. The massaman curry recipe was thought to probably be conveyed through the Bunnag family, which is the family of the Thanphuying Plian (Thompson 2002; Wongcha-um 2009). However, the name massaman (*mut-sa-mun*, มัสมั่น) did not exist in Persian or Indian language, so the word was possibly originated from the reference to Muslim who introduced the spices and curry concept to Siam (Wongthes 2017).

Other *Khaek*-influenced dishes include *khao-buri* (ข้าวบุหรี่), which is a rice dish called Kabuli believed to originate from the city of Kabul. This rice dish contains butter, milk, pepper, coriander seeds, cumin seeds, cardamom, and clove. It also includes a dessert called *Kanom Muscat* (*ka-nom-mut-sa-god*, ขนมมัศกอด), a name derived from the port city of Muscat. This baked dessert contains eggs, sugar, and flour and is topped with colored meringue. However, the only color recommended in Thanphuying Plian's cookbook is pink color from sappanwood (Chuangpichit 2001; Phassakorawong 1910). The sappanwood (*fhang*, ฝาง) is a local Thai plant, which gives color ranging from yellow to red, the pink starts at pH above seven and gets to a more intense red at higher pH (Sinsawasdi, Kwanbunjan, and Hsu 2014).

Muslim traders also brought the concept of milk-based curries, which Thai people adapted by replacing Indian butter with coconut cream. Another remarkable adaptation was to reduce the variety and number of spices in curry and add more fresh herbs (Seubsman et al. 2009).

2.5 Chinese Influence

While *Fa-rang* and *Khaek* people were regarded as foreigners, *Chin* (จีน), which is the Thai word for Chinese, were not categorized as such. The main reason might be due to the fact that Chinese people started to travel and settle in Siam about 400 years earlier than other nationals. Chinese became more influential in the 15th century when Ayutthaya took part in the Chinese trading system. With its long history, Chinese cultures began to interweave with Thai customs. Various Chinese dishes, kitchen instruments, and cooking techniques, such as noodles (*guay-teow*, ก๋วยเตี๋ยว), soy sauce (*see-ew*, ซีอิ๊ว), woks (*ga-ta-gon-lueg*, กะทะก้นลึก), wok spatulas (*ta-lew*, ตะหลิว), and stir-frying (*pud*, ผัด), were introduced by immigrants and, eventually, became part of Thai custom (Auethavornpipat 2011; Sher 2019).

Metal cookware can withstand much higher heat than the unglazed terracotta. While the terracotta pan can only be used primarily for boiling, heating metal wok with oil can reach a much higher temperature. This Chinese-influenced cooking technique allowed for the development of foods with different textures. Deep fat frying results in crispy food. Stir-frying with oil in high heat cooks food faster while retaining the firmness of vegetables and starchy ingredients like rice and noodle. As a result, Chinese-influence noodle recipes were developed in many regions.

Chanthaburi (จันทบุรี) province was famous for its high-quality rice noodle (*sen-chan* noodle, ก๋วยเตี๋ยวเส้นจันท์). The *sen-chan* noodle has the right balance of chewiness and firmness, and the noodle strips are not easily broken or shorten during the vigorous stir-frying process. The key element for this desirable property is the water from a particular source in Chanthaburi, which gives the right balance of acidity and minerals (Ministry of Culture Chantaburi 2018). Because broken rice grains were used to make this noodles, the cost of the noodle was cheaper than the staple whole-grain rice. During the economic hardship after World War II, around 1944, the Thai government encourages the shift to eat more noodles and use local ingredients, especially Thai seasoning, i.e., fish sauce, palm sugar, and tamarind paste (Puaksom 2017). Other ingredients include egg, shrimps, dried shrimps, dried salted turnip, peanuts, dried chili, and garlic chives, all available locally. The generic stir-fried noodle (*guay-teow-pad*, ก๋วยเตี๋ยวผัด) then became known as Thai-style stir-fried noodle (*guay-teow-pad-thai*, ก๋วยเตี๋ยวผัดไทย) for the additional unique tastes of hotness and sourness and locally sourced ingredients. Then, ultimately the name was shortened to only pad thai (Khowiboonchai 2013; Laulamai 2018).

Other types of Chinese-influenced stir-fried noodles are, for example, stir-fried noodle with soy sauce (*guay-teow-pad-see-ew*, ก๋วยเตี๋ยวผัดซีอิ๊ว) and stir-fried noodle with chicken (*guay-teow-kua-gai*, ก๋วยเตี๋ยวคั่วไก่) but they do not incorporate as many Thai ingredients as the pad thai. The Chinese technique of stir-frying with wok was also extended to combine with other national cuisines. For example, stir-fried macaroni (*pad-macaroni*, ผัดมักกะโรนี) is stir-fried pasta with egg, meat, and tomato sauce. *American fried rice*, which is unknown to American people, is rice stir-fried with raisins and tomato sauce, topped with fried egg, and served with fried chicken and fried frankfurter on the side of the same plate.

Nowadays, at around 14% of the Thai population, the Chinese are the largest minority group in Thailand (Auethavornpipat 2011). Therefore, the blending of Thai and Chinese cuisines and techniques has been relatively seamless. Thai-Chinese families and their recipes have been credited for establishing a street food culture that has become a significant part of Southeast Asian culinary heritage (Phongpheng, 2021). Pad Thai, is an example of Thai national culinary brand identity influenced by Chinese culture. Iconic Thai street food, which has also become the Thai culinary brand identity and signature experience for tourists, is believed to have originated in the Chinatown district of Bangkok or *Yao-wa-raj* (เยาวราช).

2.6 The Royal Influence

In the early days of Bangkok, Thai people have added snacks to our food categories. Several novels in this period captured the art and craft of how Thai foods were prepared, presented, and consumed. However, the most notable delicacies of Thai cuisine were from the royal court kitchen. Hence, these sophisticated dishes have been known as the

Royal Court Cuisine (*ar-harn-chow-wung*, อาหารชาววัง) or the cuisine of the royal household (Wongcha-um 2009).

One of the strategies used to avoid colonialism from the Western countries was to publish cookbooks as a part of becoming civilized. Thus, many cookbooks were initiated during the reign of King Rama V (which started in 1890). The first recipes were published in Pratitinbatr Journal, written by Thanphuying Plian Phassakorawong in 1888. Then, recipes were compiled and published as a book titled *Mae-Krua-Hua-Pa* in 1910. Around the same time, a cookbook titled *Tamra-Gub-Kao* by Mom Somcheen Rachanuprapan came out in 1892, then, *Tum-rub-Sai-Yaowapa* (ตำรับสายเยาวภา) in 1937 and *Tumra-Gub-Kao-Sorn-Loog-Larn* by Thanphuying Kleeb Mahidhorn in 1951. These authors were all ladies of the title who were members of the royal family or acquired their culinary skills from the royal court (Bongsanid 1935; Kasempolkoon 2017; Mahidhorn 1949; Namwong and Pewporchai 2019; Phassakorawong 1910; Rachanuprapan 1890).

The world-famous *tom-yum-koong* (spicy soup with shrimp, ต้มยำกุ้ง) was not found in any of the aforementioned cookbooks. There was *tom-yum* (spicy soup, ต้มยำ) in Mom Somcheen's cookbook, such as snakehead fish *tom-yum* (*tom-yum-pla-chon*, ต้มยำปลาช่อน) and mushroom *tom-yum* (*tom-yum-hed-kone*, ต้มยำเห็ดโคน) (Rachanuprapan 1890). However, the ingredients mentioned, e.g., garlic pickle, coriander root, and fried garlic, do not exist in the *tom-tum-koong* in a contemporary context. Another cookbook by Thanphuying Plian mentioned *Khmer tom-yum* (*tom-yum-ka-men*, ต้มยำเขมร) that is identical to outside-of-a-pot soup (*gaeng-nog-mhor*, แกงนอกหม้อ). Also, the ingredients, such as cucumber and green mango, were not found in today's *tom-yum-koong* (Phassakorawong 1910). This *gaeng-nog-mhor* recipe was recreated for this current book. Details can be found in topic 7.2. Kongpan (2018) suggested that the *tom-yum-koong* as we know it today was originated in *Klai-Kang-Won* Palace, a palace by Hua-Hin beach, in 1964. The recipe was first published in 1966 by Mom Ratchawongse Kitinuddha Kitiyakara in a book titled *Food of the Royal* (*kong-sa-veoy*, ของเสวย).

Initially, the royal court cuisine was limited to only the inner circle of those people with personal connections with royalty, the nobility, or their ladies-in-waiting. However, it has gained public interest and is known as *ar-harn-chao-wang* (อาหารชาววัง). Nowadays, Royal cuisine recipes have become more common and widespread and have blended with regular family recipes. Mom Rajawongse Thanadsri Svasti (หม่อมราชวงศ์ถนัดศรี สวัสดิวัตน์) was a famous food critique and a pioneer of gastronomic trends in Thailand and was honored as National Artist of Performing Arts in 2008. He was a member of the Thai Royal Family with the royal title of "mom rajawongse" and was raised in a Thai palace as his mother ran the kitchen for Queen Saovabha Phongsri. However, his view of the royal court cuisine is different. According to Mom Rajawonge Thanadsri, the royal court food was simply aimed at aged kings and thus needed to be easier to chew and digest. So, extra steps are to remove all inedible parts from all dishes, such as crust, fish bones, and bones. He also confirmed that royal food was not meant to be sweeter than regular food as most people believe (Bagnkok Post 2019; Bungphol 2009).

The term royal court cuisine (อาหารชาววัง) has been used in many restaurants, street food, and several cookbooks. The term added value to the business because it signaled that the food fits for a queen and delicious. In addition, the term royal court cuisine is often used as a reference standard and proof of authenticity for recipes and flavor profiles. In the view of Sounsamut (2018), the royal court-style food reflected the character of Thai people who liked to "play" (เล่น), or recreationally experimenting with food. In this regard, Thai people were known for their open-mindedness to new foods. In other words, Thai people were not prone to food neophobia.

On the contrary, Thai people embraced new ingredients brought along by foreigners. They had fun experimenting and blending the local elements with imported ones to create new dishes. As a result, newly invented dishes or fusion food at the time have proliferated and gradually shaped the identity of Thai cuisine.

If playing or experimenting with ingredients is the basis of Thai cuisine, the food culture by itself is a better representation of Thai cuisine than a fixed or literal recipe. This sense of experimentation, the attention to food details with dedication, especially from those who worked for the royal family, along with the intelligence of the royal family members who transferred the knowledge in writing, are the basis of innovation into the royal cuisine (Kongpan 2020). Generally, the main characteristics of the royal cuisine are well-balanced flavors (taste and aroma), high-quality ingredients, and captivating presentation (Kasempolkoon 2017).

With the modern lifestyle, the royal court still plays a substantial role in the world of Thai cuisine, though not as a center of food innovation or a trendsetter. Her Royal Highness, Princess Maha Chakri Sirindhorn, still enjoying experimental food and publishing her work. She also initiated the Royal Traditional Thai Crafts School for Women located inside the Royal Palace. The school transfers the knowledge and skills of arts and crafts, especially cooking and vegetable craving, to the new generations to transfer the artistic schools to the new generations and protect and preserve the sophisticated Thai culinary heritage (Royal Office 2021).

2.7 The Contemporary Influence of Government Policy

In another aspect of Thai cuisine, the Thai government has recognized the worldwide market potential of Thai food. There are many gastro-diplomacy campaigns. In an attempt to increase the country's income from gastronomy tourism, the Tourism Authority of Thailand (TAT) promoted Thai food through the "Amazing Thailand campaign in 1998, which helped to rebrand Thai food from mere exotic foods to the premier city for world-class cuisines (Phongpheng, 2015).

Another national project is Thai Kitchen to The World (Government Public Relations Department Thailand n.d.; Karp 2018; Lam 2019; Suntikul 2019). This project included the Thai Delicious, initiated by the Ministry of Science and Technology and managed by the National Innovation Agency (NIA). The main objective of Thai Delicious was to standardize Thai food tastes and recipes. Standardization may be helpful. For example, in the past, some Thai restaurants abroad can be criticized for too much modification and were accused of "corrupting real Thai cuisine" (Padoongpatt 2017).

The NIA successfully introduced an analysis instrument called "e-delicious." The device composes of an "electronic-nose" with 16 gas sensors to detect aroma compounds, an "electronic-tongue" to analyze the five basic tastes (sweetness, sourness, saltiness, bitterness, and umami), and a central processing unit to interpret results (National Innovation Agency Thailand 2014; Varanyanond 2013). The Thai Delicious project included a certification service for a restaurant dish that met the established *Thai Delicious* standards. The judgment was performed by the ESenS instrument (second generation of the e-delicious). In one example shown to the news reporters, from the highest score of 100%, the family recipe *tom-yum-koong* of the famous Jay Fai's Michelin-starred restaurant in Bangkok earned a score of 90%, meaning the food met the standard. This procedure aimed to ensure that every Thai dish, such as pad thai and green curry, tastes the same in any restaurant around the world. The concept made big headlines worldwide in 2014. (Bangkok Biz News 2014; BBC News 2014; Head 2014; Pajee 2014). Those headlines included "Government Robots Will Decide if Your Thai Food Tastes Right, Thai government-created robots will evaluate Thai food and give it a grade" by ABC News (Dewast 2014).

Later in 2017, TAT collaborated with food certification firm Michelin Guide to initiate the first Michelin guidebook. The guidebook has attracted both local and international attention to restaurants and street food stalls (Sritama, 2017). Another movement to guarantee authentic Thai taste is from the Ministry of Commerce, the Royal Thai Government. The ministry awards the "Thai SELECT" mark of certification to Thai ready-to-eat/ready-to-cook products and Thai restaurants in Thailand and overseas. Details are listed on www.thaiselect.com.

Thai cuisine has been depicted internationally through time, from the Thai Royal Court banquet in the 1860s, to the state campaign such as "Thai Kitchen to the world" in 2010. The government involvement is perceived as culinary colonialism, described as the mobilization of state power through food. Professor Penny Van Esterik (2018), an anthropologist with long experience in Thailand, stated that "the Thai state always was and remains extremely conscious of what others think of the nation. No colonial experiences have convinced the Thai that other countries, or other cuisines are 'better' than theirs."

2.8 Conclusion

The gastronomy development of Thai cuisine has been very dynamic. The world-famous dishes, such as *tom-yum-koong* and pad thai, though they were honored as Thai national heritage, have just been developed in the last century. The constant changes and development in Thai cuisine may be because Thai people enthusiastically embraced and elaborated ingredients and techniques from other food cultures. Profound food cultures incorporated in Thai cuisine were simplified into *Fa-rang*, *Khaek*, and *Chin*. The center of culinary expertise was evident in the royal court kitchen. Then, with the publication of cookbooks, the exclusive recipes once considered innovations have been transferred to and enjoyed by the general public. However, the excessive modification of Thai dishes, especially from Thai restaurants abroad, prompted the Thai government to standardize Thai food recipes, food products and food services that meet the criteria get award and recognition, an attempt to encourage authenticity of the Thai cuisine.

References

Auethavornpipat, Ruji. 2011. "Flexible identity: Unfolding the identity of the Chinese-Thai population in contemporary Thailand." *The Arbutus Review* 2 (1):32–48.

Bagnkok Post. 2019. *National Artist MR Thanadsri Svasti Passes Away; Remembered for Talents, Gastronomic Passion*. Bangkok, Thailand: Bangkok Post.

Bangkok Biz News. 2014. "Delicious Measuring Device for Thai Kitchen (เครื่องวัดความอร่อยครัวไทย)." In *Bangkok Biz News* (กรุงเทพธุรกิจ). Bangkok, Thailand: Bangkok Biz News. https://www.bangkokpost.com/life/arts-and-entertainment/1737783/national-artist-mr-thanadsri-svasti-passes-away.

BBC News. 2014. Thailand's Ex-PM Develops Food Robot to Test Thai Food. https://www.bbc.com/news/technology-29424311.

Bongsanid, Yaovabha. 1935. *Tumrub Sai Yaowapa* (ตำรับสายเยาวภา). Bangkok, Thailand: Saipunya Sakakom.

Bruce, Robert. "KING MONGKUT OF SIAM AND HIS TREATY WITH BRITAIN." *Journal of the Hong Kong Branch of the Royal Asiatic Society* 9 (1969): 82–100. http://www.jstor.org/stable/23881479.

Bungphol, Thanok. 2009. "Tales of Royal Court Cuisine (เล่าขานอาหารชาววัง ใต้เบื้องพระยุคลบาท)." *Matichon Weekly*, 27 February 2009, 27.

Castillo, Cristina Cobo, Bérénice Bellina, and Dorian Q. Fuller. 2016. "Rice, Beans and Trade Crops on the Early Maritime Silk Route in Southeast Asia." *Antiquity* 90 (353):1255–1269. doi: 10.15184/aqy.2016.175.

Castillo, Cristina Cobo, Katsunori Tanaka, Yo-Ichiro Sato, Ryuji Ishikawa, Bérénice Bellina, Charles Higham, Nigel Chang, Rabi Mohanty, Mukund Kajale, and Dorian Q. Fuller. 2016. "Archaeogenetic Study of Prehistoric Rice Remains from Thailand and India: Evidence of Early Japonica in South and Southeast Asia." *Archaeological and Anthropological Sciences* 8 (3):523–543. doi: 10.1007/s12520-015-0236-5.

Chuangpichit, Teeranun. 2001. "ตามรอย สำรับแขกคลองบางหลวง (Khaeg Cuisines in Klong Bang Luang)." *Sarakadee.*

Dewast, Louise. 2014. Government Robots Will Decide if Your Thai Food Tastes Right.

Dhiravegin, Likhit. 2002. การเมือง การ ปกครอง ของ ไทย. Thammasat Press สำนัก พิมพ์ มหาวิทยาลัย ธรรมศาสตร์.

Government Public Relations Department Thailand. n.d. Thailand: Kitchen of the World. https://thailand.prd.go.th/ebook/B0028/index.html#p=1.

Head, Jonathan. 2014. *Thai Tasting Robot tastes for Authenticity.* Bangkok, Thailand: BBC News. https://www.bbc.com/news/world-asia-30110282.

Kanchanakunjara, Taddara, Songkoon Chantachon, Marisa Koseyayothin, and Tiwatt Kuljanabhagavad. 2015. "Traditional Curry Pastes During Sukhothai to Ratthanakosin: The Subjective Experience of the Past and Present." *Asian Culture and History* 7 (1):175.

Karp, Myles. 2018. "The Surprising Reason that There Are So Many Thai Restaurants in America." @vice. https://www.vice.com/en/article/paxadz/the-surprising-reason-that-there-are-so-many-thai-restaurants-in-america.

Kasempolkoon, Aphilak. 2017. "When "Royal Recipes" Became More Widely Known: Origin and Development of "Royal Cookbooks" in King Rama V's Reign to Girls' Schools' Establishment." *Vannavidas* 17:353–385. doi: https://so06.tci-thaijo.org/index.php/VANNAVIDAS/article/view/107539.

Khowiboonchai, Poonpon. 2013. "Power negotiation and the Changing Meaning of Pad Thai: From Nationalist Menu to Popular Thai National Dish." *Journal of Language and Culture* 32 (2):75–75.

Kim, Kyung-Joong, and Chang-Ho Chung. 2016. "Tell Me What You Eat, and I Will Tell You Where You Come From: A Data Science Approach for Global Recipe Data on the Web." *IEEE Access* 4:8199–8211. doi: 10.1109/ACCESS.2016.2600699.

Kongpan, Srisamorn. 2018. *Intangible Cultural Heritage Foods of Thailand* (อาหาร ขึ้นทะเบียน มรดกทางภูมิปัญญาของชาติ). Bangkok, Thailand: S.S.S.S. (บริษัท ส.ส.ส.ส. จำกัด).

Kongpan, Srisamorn. 2020. "Brief Biography of Famous Lady Chefs of the Palaces (เล่าสั้นๆ)." In *The Story of Thai Food (*เรื่องเล่ากับข้าวไทย*)*, 124–125. Bangkok, Thailand: S.S.S.S.

Lam, Francis. 2019. "How Thai food took over America." The Splendid Table. https://www.splendidtable.org/story/2019/01/10/how-thai-food-took-over-america.

Laulamai, Krit. 2018. "Pud (Mai) Thai." In *O-Cha-Ka-Le*, 67–74. Bangkok, Thailand: Way of Book.

Mahidhorn, Thanphuying Kleeb. 1949. "Cookbook for Offsprings (หนังสือกับข้าว สอนลูกหลาน)." *Vajirayana Digital Library.* https://vajirayana.org. https://vajirayana.org/

Ministry of Culture Chantaburi. 2018. Chantaburi Rice Noodle (ก๋วยเตี๋ยว เส้นจันท์). https://www.m-culture.go.th/chanthaburi/ewt_news.php?nid=932&filename=index.

Namwong, Jutarat, Pewporchai Passapong. 2019. "Causative Verbs of Cooking: Study from Recipe of Mom Som Cheen Rachanuprapan 1892 (คำ กริยา กา รี ต เกี่ยว กับ การ ประกอบอาหาร: ศึกษา จาก ตำรา กับ เข้า ของ หม่อม ซ่ ม จีน ราชา นุ ประพันธ์ ร. ศ. 110)." Conference Proceedings, 10th Global Goals, Local Actions: Looking Back and Moving Forward (รายงานการประชุม วิชาการ เสนอ ผล งาน วิจัย ระดับ ชาติ และ นานาชาติ), 1(10), 699–711. Suan Sunandha Rajabhat University, Bangkok, Thailand.

National Innovation Agency Thailand. 2014. *Thai Delicious* (ศาสตร์แห่งรสชาติอาหาร ไทย). Edited by Supachai Lorlowhakarn. Thailand National Innovation Agency. https://nia.bookcaze.com/viewer/1262/1.

Nitivorakarn, Saruda 2014. "Thai Food: Cultural Heritage of the Nation." *Academic Journal Phranakhon Rajabhat University* 5 (1):171–179.

Padoongpatt, Mark. 2017. "3. Too Hot to Handle? Restaurants and Thai American Identity." In *Flavors of Empire*, 85–117. California: University of California Press.

Pajee, Parinyaporn. 2014. Standards for the Thai kitchen. Accessed 2014-10-06.

Pewporchai, Passapong. 2017. "A study of Cooking Terms in Thai Recipe Book: A Case of Her Ladyship Plain Phassakorawong's "Mae Krua Hua Pa" Recipe Book." *Journal of Liberal Arts, Ubon Ratchathani University* 13 (2):138–165.

Phassakorawong, Thanphuying Plian. 1910. *Tamra maekhrua hua pa*. Edited by Plain Phassakorawong. Bangkok: Tonchabap. https://vajirayana.org.

Phongpheng, Vannaporn 2015. *"BANGKOK AS A 'METROPOLIS OF EXOTIC CUISINES': THE EPRESENTATION OF 'CIVILIZED BANGKOK' FROM FOOD CULTURE."* Ph.D. thesis, Bangkok, Thailand: Chulalongkorn University, p. 220–296.

Phongpheng, Vannaporn. 2021. Food History: A Case of Bangkok Street Food Stalls in the Michelin Guide. SSRN 3957363: paper is partially funded by The Empowering Network for International Thai and ASEAN

Puaksom, Davisakd. 2017. "National Food Policy 1940 (โครงการส่งเสริมอาหารของชาติ ยุคจอมพล ป. 2481 'บันไดขั้นแรกของการสร้างชาติ')." *Hfocus*. https://www.hfocus.org/content/2017/07/14292.

Rachanuprapan, Mom Somcheen 1890. *Tum-ra-gub-kao (*ตำรากับเข้า*)*. Bangkok, Thailand: Watcharin.

Royal Office. 2021 Royal Traditional Thai Crafts School for Women. https://www.royaloffice.th/en/home/.

Seubsman, Sam-ang, Pangsap Suttinan, Jane Dixon, and Cathy Banwell. 2009. "20 – Thai Meals." In *Meals in Science and Practice*, edited by Herbert L. Meiselman, 413–451. Cambridge, UK: Woodhead Publishing.

Sher, Israel. 2019. "Clash of "Chailand"; The Chinese Influence on the Thai Cuisine." The University of Gastronomic Sciences of Pollenzo, Italy, Last Modified 2 Novermber 2019. https://thenewgastronome.com/clash-of-chailand/.

Sinsawasdi, Valeeratana K., Karunee Kwanbunjan, and Wei-Yea Hsu. 2014. "Sensory Expectation and Perception of Red Bverages Prepared from sappanwood (*Caesalpinia sappan* L.) Water Extract." *Proceedings of the Biodiversity Biotechnology Bioeconomy, Thailand* 26:388–393.

Sinsawasdi, Valeeratana Kalani, Chutamas Jayuutdiskul, Prangmilint Montriwat, and Karunee Kwanbunjan. 2020. "Milk Buying Decision for Preadolescents Reflects Societal Value of Being Tall: A Focus-Group Study of Parents in Bangkok." *Journal of International Food & Agribusiness Marketing* 34 (1):77–95.

Sounsamut, Pram. 2018. "ตำรับ ชาววัง: นวัตกรรม แห่ง โภช ณี ย ประณีต (Tam Rap Chao Wang: Innovation of the Food Refinement)." *The Journal of Social Communication Innovation (วารสาร วิชาการ นวัตกรรม สื่อสาร สังคม)* 5 (2):150–160.

Srisophon, Thongchai. 2018. *History of Thailand* (ประวัติศาสตร์ ชาติ ไทย). Edited by กลุ่ม วิชา ประวัติศาสตร์. Sounsamut, Pram: Buriram Rajabhat University Press.

Sritama, Suchat. 2017. "Michelin Guide set to hit Thai tables; Star system to light up food hub prospects." *Bangkok Post*, 29 November 2017, Business. https://www.bangkokpost.com/business/1368711/michelin-guide-set-to-hit-thai-tables.

Sujachaya, Sukanya. 2017. "Thai Cuisine during the Ayutthaya Period." *Humanities Journal* 24 (2): 1–29. doi: https://www.tci-thaijo.org/index.php/abc/article/view/106404.

Sukphisit, Suthon. 2019. "Chilli's Complicated History; Where Did This Most Indispensable Thai Cooking Spice Originate?" *Bangkok Post*, 5 May 2019, B Magazine. https://www.bangkokpost.com/life/social-and-lifestyle/1672304/chillis-complicated-history.

Suntikul, Wantanee. 2019. "Gastrodiplomacy in Tourism." *Current Issues in Tourism* 22 (9):1076–1094. doi: 10.1080/13683500.2017.1363723.

Suppakul, Panuwat, Thitiporn Thanathammathorn, Ornsiri Samerasut, and Surachai Khankaew. 2016. "Shelf life Extension of 'fios de ovos,' An Intermediate-Moisture Egg-Based Dessert, by Active and Modified Atmosphere Packaging." *Food Control* 70:58–63. doi: https://doi.org/10.1016/j.foodcont.2016.05.036.

Teanpewroj, Petchrung. 2017. "Diversity of Food Culture in Ayutthaya Period ความ หลากหลาย ของ วัฒนธรรม อาหาร การ กิน สมัย อยุธยา" *Songklanakarin Journal of Social Sciences and Humanities* 23 (1): 67–89.

Thompson, David. 2002. "Mussaman Curry of Chicken." In *Thai Food*, 329–332. London, UK: Pavillion Books.

UNESCO. 2021. Did You Know?: Thailand and the Maritime Silk Roads | Silk Roads Programme. https://en.unesco.org/silkroad/content/did-you-know-thailand-and-maritime-silk-roads.

Van Esterik, Penny. 2018. "Culinary Colonialism and Thai cuisine: The Taste of Crypto-Colonial Power." Dublin Gastronomy Symposium. Food and Power., Dublin, Ireland.

Varanyanond, Warunee. 2013. *Fostering Food Culture with Innovation: OTOP and Thai Kitchen to the World*. Bangkok: Institute of Food Research and Product Development, Kasetsart University. Retrieved March 9:2016.

Weber, Steve, Heather Lehman, Timothy Barela, Sean Hawks, and David Harriman. 2010. "Rice or Millets: Early Farming Strategies in Prehistoric Central Thailand." *Archaeological and Anthropological Sciences* 2 (2):79–88. doi: 10.1007/s12520-010-0030-3.

Wongcha-um, Panu. 2009. What is Thai Cuisine? Thai Culinary Identity Construction from the Rise of the Bangkok Dynasty to Its Revival. http://scholarbank.nus.edu.sg/handle/10635/17685.

Wongthes, Sujit. 2017. *Where Did Thai Food Come From?* (อาหารไทยมาจากไหน). Bangkok, Thailand: Natahak.

Yankowski, Andrea, Puangtip Kerdsap, and Nigel Chang. 2015. ""Please Pass the Salt" – An Ethnoarchaeological Study of Salt and Salt Fermented Fish Production, Use and Trade in Northeast Thailand." *Journal of Indo-Pacific Archaeology* 37:4–13.

Yeh, S. D., H. J. Bau, Y. J. Kung, and T. A. Yu. 2007. "Papaya." In *Transgenic Crops V (Biotechnology in agriculture & forestry, Vol. 60)*, edited by Michael R. Davey Eng-Chong Pua, 73-96. Springer Berlin, Heidelberg.

PART II

Multisensory Properties of Thai Foods and Their Sources

3

Flavor Components and Their Sources in Thai Cuisine

VALEERATANA K. SINSAWASDI AND NITHIYA RATTANAPANONE

Contents

3.1 Fundamentals of Food-Sensory Characteristics

Each Thai dish can be very complicated, which can be related to several disciplines in science, including psychological, nutritional, social, historical, biological, cultural, and foodservice industry perspectives (Meiselman 2009). The sensory dimensions of food involve the perceptions through the nose and mouth, i.e., the chemical senses of taste and smell. In addition, the spatial senses of touch, texture,

DOI: 10.1201/9781003182924-5

temperature, pain, hearing, and vision also influence perceived sensations of food (Spence 2020).

Our pleasure in eating and the whole eating experience lead to appetite and preference for particular foods and cuisines. Thus, the total experience during food consumption is complex and can be regarded as responses to or perceptions of stimuli (foods and drinks). Perception of the flavor and taste of food involve stimulating several senses: the gustatory sense (taste), the olfactory sense (smell), the trigeminal sensations that is stimulated by irritants (pungency, cooling, numbing, tingling), the tactile sense or the sensing of texture (mouthfeel), and the sense of sight and hearing. We perceive food as a combination and interaction of these taste, smell, vision, hearing, and touch modalities (Rogers and Blundell 1990; Stubbs et al. 2001; Yeomans, Weinberg, and James 2005; Lawless and Heymann 2010).

To the eyes of a foreigner who did not grow up with Thai food culture, the sensory perception of Thai cuisine has been described as tasty, aromatic, looking pleasing, and attractive (Jang, Ha, and Silkes 2009). Thus, Thai foods are stimuli that can pleasantly evoke all human senses.

3.2 Tastes

There are a limited number of perceivable tastes compared to perceivable odors. The concept of basic taste has been studied throughout history. Early evidence led scientists to conclude that there are four basic tastes: sweet, salty, sour, and bitter (McBurney and Gent 1979). From the late 1800s, it was believed that each region of the tongue senses a specific taste. This theory was false but somehow managed to exist in many books decades after being proven wrong in the 1970s (Smith and Margolskee 2001). With the advance in molecular biology and pieces of evidence on specific taste receptors, more basic tastes have been proposed. Since confirming a specific glutamate taste receptor (taste-mGluR4), umami has been accepted as the fifth basic taste. All five basic tastes, i.e., sweet, salty, sour, bitter, and umami, can be perceived through the taste receptors on our tongues (Kurihara 2009).

In Thai culinary culture, the phrases used to describe food tastes are "sour-sweet-fatty-salty" (*preow-waan-mun-kem*, เปรี้ยวหวานมันเค็ม), "delicious" (*a-roy*, อร่อย), and "well-rounded" (*gloam-glom*, กลมกล่อม).

There is also a phrase with an element of pungency, such as sour-sweet-fatty-salty-hot (*preow-wan-mun-kem-ped*). Less frequently used terms such as *fhard* (ฝาด) and *zah* (ซ่า) refer to the astringency, numbness, tingling, and fizzy sensations felt in the mouth. The ingredients serving as the primary sources of sweet, sour, and bitter tastes are briefly listed in this chapter.

Umami is discussed in the most detail as it significantly contributes to the overall taste and preference for Thai cuisine. However, tasting each ingredient by itself does not always provide the umami taste. A relatively newly discovered taste is a savory sensation called kokumi, which is also mentioned briefly.

Lately, several studies have supported speculation that our taste receptors might be able to sense fatty taste, especially when free fatty acid is present. The proposed name for this new taste is oleogustus. However, the research on this sixth basic taste so far has been on its relationship with obesity (Keast and Costanzo 2015; Running, Craig, and Mattes 2015; Running and Mattes 2016; Kure Liu et al. 2019). There is not enough linkage to food-sensory perception, so this book has not included the fat sense.

Hotness, piquancy, or pungency is not included in basic taste. This burning sensation from chili pepper has been proven to be perceived through the trigeminal nerve and is thus not considered a flavor. So, chili peppers will be discussed as irritant stimuli to the trigeminal system.

3.2.1 *Common Sources of the Four Basic Tastes*

The dominant seasoning for the salty taste in Thai food is fish sauce. Other salty seasonings include *pla-ra* (ปลาร้า), *budu* (บูดู), *tua-nao* (ถั่วเน่า), *nam-poo* (น้ำปู๋), soy sauce, and salt. Details of these seasonings are in Chapter 4 under the topic of fish sauce and other salty seasonings and in this chapter under the topic of umami.

The main seasoning for sourness is lime juice. However, when limes are out of season, other fruits can be used for sourness. Examples are kaffir lime juice, bitter orange (*som-sa*, ส้มซ่า), bilimbi (*ta-ling-pling*, ตะลิงปลิง), hairy-fruited eggplant (*ma-aueg*, มะอึก), sour green mango (*ma-muang-preow-dib*, มะม่วงเปรี้ยวดิบ), sour tamarind juice (*nam-ma-karm-piag*, น้ำมะขามเปียก), garcenia (*som-kaeg*, ส้มแขก), sour garcinia (*ma-dun*,

มะดัน), wild tomato (*ma-kuae-som*, มะเขือส้ม), and salak (*sa-la*, สละ). Many recipes call for a combination of these sour ingredients. As Thai cuisine is a loose approach, cooks can rely on their senses to judge the final quantity of these ingredients.

Nowadays, crystallized sucrose from sugarcane is the most popular form of sweetener. Traditionally, syrup from the flowers of the palm tree or coconut tree has been used as sweeteners. The syrup can be heated to achieve a higher concentration of sugar and extend its shelf life. This type of sugar is brown and has a sweet floral aroma. In fact, the word sugar in Thai is the word for brown color (*nam-tarn*, น้ำตาล). Palm sugar from the flower of the palm tree (*Borassus flabellifer*) is called *nam-tarn-ta-node* (น้ำตาลโตนด). Sugar from the coconut palm (*Cocos nucifera*) is called *nam-tarn-ma-prao* (น้ำตาลมะพร้าว). More details on this palm sugar are in Chapter 4.

Bitterness is desirable only in certain dishes. There is no seasoning to add to a bitter taste, and the well-known source of bitterness is the bitter melon (*ma-ra*, มะระ).

3.2.2 Umami, the Fifth Basic Taste

Umami is a pleasant savory taste. There is no direct translation in English, so terms such as brothy, meaty, and savory have been used. There is no direct translation of umami in the Thai language either but it can be understood vaguely as brothy (as in the sweetness of bone broth) or the universal term *a-roy* (อร่อย) for being delicious, and an obsolete term such deliciousness of meat (*o-cha*, โอชะของเนื้อสัตว์).

The deliciousness of umami is enhanced by mixing ingredients rich in glutamate and nucleotides that have been traditional in many world cuisines and cultures. For example, the famous Japanese dashi soup, which is how umami was discovered as a perceivable taste, is a combination of glutamate from konbu (kelp seaweed) and nucleotides from bonito (dried skipjack tuna flakes) (Kurihara 2009; Marcus 2009). Thus, any food that contains this combination of glutamate and nucleotides also exhibits the umami taste characteristic. Culinary terms used by chefs to describe umami are, for example, "yummy," "umami synergy," and "u-bombs."

In the form of free glutamic acid, glutamate is a nonessential amino acid and a natural component of several foods such as meat, poultry, tomatoes, soybean, and Parmesan cheese. In addition, there is a natural glutamate present in human saliva and even in human breast milk. This glutamate content in breast milk is as high as in Japanese dashi soup leading to the belief that infant exposure to the high umami content in breast milk familiarizes us with the savory taste of umami from the beginning of life (Yamaguchi and Ninomiya 2000).

Various types of nucleotides are found naturally, especially in animal products. In general, meats contain nucleotides in the form of 5′-inosinate (IMP). Fish and shellfish have 5′-adenylate (AMP). In food preservation processes such as fermentation and drying, protein molecules break down into several smaller derivatives. Thus, bonito or dried fermented tuna fish has a particularly high AMP and IMP content. Nucleotides are also found in plants, especially when undergoing the drying process. Notable examples are dried shiitake mushrooms containing 5′-guanylate (GMP) (Giacometti et al. 1979; Ninomiya 1998).

Combining glutamate and nucleotides has a synergistic effect, which means the umami taste is greatly enhanced. In other words, the same level of umami intensity can be achieved with a lesser amount of glutamate if there are nucleotides in the mixture (Kuninaka 1960).

3.2.2.1 Ingredients with Umami Taste Substances   The umami characteristics in Western cuisine also achieve an umami taste from this synergy. French meat stock contain glutamate. French onion soup has both glutamate from meat stock and IMP from Parmesan cheese. Italian foods, such as the bolognese spaghetti sauce, combine glutamate from tomato with glutamate and IMP from meat. American cheeseburger tastes better with cheddar cheese that contains glutamate and IMP from the grilled beef patty (Kurihara 2009; Marcus 2009).

For Thai cuisine, similar to many other Asian cuisines, seasonings rich in free glutamate, such as soy sauce and fish sauce, have been used to enhance umami and improve the overall taste of many dishes. Most of these seasonings come from the drying and fermentation of seafood, beans and grains, and salt. The process yields umami substances, mainly glutamate. These seasonings made from fermentation are used as sources of saltiness and the umami taste throughout Asian cuisine, including Thailand. Dried fish and mushrooms are used in Korea, Japan, and China; fermented soybean paste, fermented shrimp, and fermented fish in Thailand, China, Japan, Korea, Malaysia, Vietnam, and the Philippines (Ruddle and Ishige 2010; Hajeb and Jinap 2015). These seasoning sauces contain glutamate at between 620 and 1,380 mg/100 g, and the meat of fish products generally used in recipes has IMP in the range of 70–285 mg/100 g. A combination of seasoning and some meat thus creates a desirable, delicious taste. When replacing salt with these seasonings, the umami develops, and saltiness is enhanced. Therefore, a lower sodium version of the dish can be achieved without compromising the palatability (Hartley, Liem, and Keast 2019).

The manufacturing process of umami seasoning usually requires either drying, fermentation, or both for the umami taste to develop. These processes enable the decomposition of protein into amino acids and nucleotides. Just like fish sauce, several other Thai fermentation products have been used as a source of the umami taste. There are both animal and plant sources of protein. Animal sources are saltwater fish, freshwater fish, shrimp, and krill, while the plant source is soybean. Bacterial cultures and enzymes for fermentation are endogenous to the raw materials and the local environment, especially halophilic bacteria that already exist in the fish's internal organs. Most of them also develop a desirable brown color with a characteristic smell.

Depending on the region, raw materials, recipes, the final product organoleptic properties, and flavors vary but are not interchangeable. In Thai cuisine, fermented fish seasoning such as *nam-pla* (fish sauce, น้ำปลา), *bu-du* (บูดู), and *pla-ra* (ปลาร้า) and fermented shrimp such as *ga-pi* (กะปิ) are used specifically in each dish. Replacing these seasonings with salt or soy sauce, enhanced with monosodium glutamate (MSG) or other Thai seasonings, will alter the character of most dishes to the point where they cannot be identified as the original.

3.2.2.2 Umami in Clear Soup   Just as the organoleptic quality of fish sauce depends largely on the hydrolysis product of fish protein from different types of fish, the flavor and consumer acceptance of bone broth is also subject to the type and amount of amino acid derived from protein in the meat or the bone. When chicken bone broth or soup stock is simply cooked by heating water and chicken bones, the umami taste emerges as free amino acids, especially glutamate and alanine, which are released due to the hydrolysis of protein (Ninomiya et al. 2010; Pérez-Palacios et al. 2017).

Visible changes during heating are the aggregation of heat coagulating soluble protein into a floating solid mass, and simple skimming or scooping out will leave a clearer broth. Another detectable change is the development of the aromatic nature of a stock character, possibly from Maillard's reaction. Soup, heated at a higher temperature and for a longer time, tends to have higher amino acid content. More amino acid, especially glutamate, contributes to a greater intensity of umami. A study on the overall liking and sensory properties such as color, flavor, clearness, and mouthfeel of chicken stock prepared at different temperatures and heating times revealed that the best flavor soup is achieved when heating at the temperature of 95°C for 3 hours (Chotechuang, Lokkhumlue, and Deetae 2018). The finding supports the recommendation of Thai-style clear soup (*gaeng-jued*, แกงจืด) to boil meat cut with bone with high heat long enough for umami taste (*o-cha* of meat, โอชะของเนื้อสัตว์) is obtained (Sanitwong 1980).

In clear soup with ingredients such as minced pork, soft tofu, and Chinese cabbage, the adding of fish sauce, even without bone broth, will yield an umami taste from the combination of the glutamate in the fish sauce and the IMP in the pork. The fish sauce contains 950 mg/100 g, and cabbage contains 50 mg/100 g of free glutamic acids, while pork has 200 mg/100 g of IMP and a little bit of GMP and AMP (2 and 9 mg/100 g, respectively). The synergistic effect between glutamate and nucleotides, especially IMP, helps to increase the intensity of umami even at very low concentrations of each (Yamaguchi and Ninomiya 2000). Compared with sourness and saltiness, the umami taste lingers in the mouth for longer, creating overall pleasure after the meal (Horio and Kawamura 1990).

There are several recipes in Thai cuisine that are based on bone broth and seasoned with fish sauce. The richness of the umami taste is the result of the synergistic effect of glutamate and nucleotides; for example, clear vegetable soup (*gaeng-jued*, แกงจืด) with various vegetables, such as cabbage (*ka-lum-plee*, กะหล่ำปลี), Chinese cabbage (*puk-kard-khao*, ผักกาดขาว), ivy gourd leaves (*tum-lueng*, ตำลึง), wax gourd (*fug-keow*, ฟักเขียว), ridge gourd (*buab*, บวบ), daikon (*hua-chai-tao*, หัวไชเท้า), bamboo shoot (*nor-mai*, หน่อไม้), bitter melon (*ma-ra*, มะระ), fermented cabbage (*pak-kard-dong-preow*, ผักกาดดองเปรี้ยว), glass noodles (mung-bean vermicelli, cellophane noodles, *woon-sen*, วุ้นเส้น), seaweed, meat, tofu, and egg; or in noodle soup with varieties of noodles and meat products.

Though there are vast varieties of *gaeng-jued*, their basic recipes are quite similar. Ingredients that need longer heat to soften, such as bitter melon or wax gourd, are added soon after the boiling starts. Softer vegetables, such as ivy gourd leaves, are added right before serving to prevent an overcooked texture. Bone stock is also an optional basis for other spicy soups such as *tom-yum*, *gaeng-liang* (แกงเลียง), and *gaeng-som* (แกงส้ม); and curry with coconut milk such as *tom-ka* (ต้มข่า), red curry (*gaeng-ped*, แกงเผ็ด), green curry (*gaeng-keow-wan*, แกงเขียวหวาน), and yellow curry (*gaeng-ga-ri*, แกงกะหรี่) (Bongsanid 1935; Xoomsai 2012). However, the stock is not as essential for enhancing palatability in spicy soup as in the case of vegetable or noodle soup because the umami taste tends to be overpowered or suppressed by the burning sensation of chili pepper, strong aroma from the spices, or astringency and sourness.

Thai fish sauce has distinctive umami even when tasted by itself as a single ingredient (Ritthiruangdej and Suwonsichon 2007). So the fish sauce can be applied to any soup and still provide deliciousness. Thai commercial fish sauce in this book refers to the protein hydrolysate solution obtained from the natural fermentation of the anchovies. There are a variety of cheaper types called "mixed fish sauce," which is the genuine one diluted down by adding water, salt, and color and flavor additives. However, the aroma and taste of the lower grade fish sauce, especially umami, are far less preferable.

The umami from glutamic acid and nucleotides in Southeast Asian cultures is surprisingly similar to the ancient fish sauce, *Allec*, produced over 2,000 years ago by the Romans. Discovering *Allec* is evidence of the comparable processing methods with fish and salt as the only two ingredients. Thus, it is the action of proteolysis enzymes in fish that hydrolyzes (breaks down) the fish protein to various flavor compounds. Evidence of large fish sauce businesses in ancient cities like Pompeii gives us reason to believe that over 2,000 years ago, Western foods may have been characterized by an umami taste, similar to today's oriental cuisine (Curtis 2009).

3.2.2.3 Role of Umami in Salt Reduction The interaction between umami and saltiness enhances the overall taste or palatability of soup.

Hence, the reduction of salt does not diminish the overall acceptability of the soup (Yamaguchi and Takahashi 1984; Chi and Chen 1992). Research into this speculation on other foods is consistent with these findings. For example, the salt reduction was not detected when 0.4% MSG was added to 40% salt reduction recipes of Singaporean-style chicken rice and spicy noodle soup (Leong et al. 2016). In another experiment, three food models were cooked: chicken broth (skinless chicken thighs/legs, water, ginger, garlic, onion, and spices), tomato sauce (tomatoes, onion, garlic, basil, spices), and coconut curry (curry powder, coconut milk, chicken broth, shallots, lemongrass, herbs, and spices). When replacing saline (sodium chloride solution) with fish sauce for up to 25% of the total sodium in chicken broth, these foods' palatability and consumer acceptance were unaffected.

For the tomato sauce and coconut curry, the fish sauce replacement was 16% and 10%, respectively (Huynh, Danhi, and Yan 2016). The reason for this variation between broth and stock is unclear. However, it helps to confirm that dishes with free glutamate seasonings (such as fish sauce) are likely to be more delicious because of the umami taste development and unique aromatic compounds. They are also likely to contain less sodium compared with those dishes seasoned only with salt.

3.2.2.4 MSG as Food Additive   Since most of the food with an umami taste is savory, salt or sodium chloride is usually added for saltiness. However, for instant enhancement of umami in the dish with lower natural glutamate or accelerating the cooking process, a commercial form of glutamate is widely added as a chemical food additive. The Ajinomoto Company first produced the product in Japan and worldwide production is about two million tons (or two billion kilograms) per year (Sano 2009). The commercial glutamate product is in the form of monosodium glutamate crystals, widely known as MSG. Thailand is among those countries with the high consumption of MSG, averaging almost 1.5 g per day, which is about the same as Japan and Korea and three times higher than average MSG consumption in the United States (Mustafa et al. 2017). Instant noodles, lacking natural glutamate and nucleotides because the main ingredient is solely noodles with no meat, rely on umami from MSG and IMP for their meaty and umami taste. Thai people consume instant noodles, an average of 50 servings per person per year. Popular instant noodle flavors for Thai people are those with hot chili pepper (WINA 2019). In addition, about half of instant noodles sold in Thailand are *tom-yum-koong* flavor (Ratirita 2018).

3.2.3 Kokumi, Possibly the Sixth Basic Taste

Recently, there has been the discovery of taste compounds that enhances the flavor of savory food and contributes to the continuity (lingering, long-lasting), mouthfulness, thickness, and complexity of the body of food. This savory taste sensation is called kokumi, and the stimuli for this taste are peptides and amino acids. The amino acids in peptides, which are responsible for the koku sensation, are glutamine, valine, and glycine (γ-Glu-Val-Gly). For those familiar with cheese, the kokumi taste is easily recognized when comparing mature Gouda cheese (with a higher kokumi taste) with a mild one. Besides cheese, other fermented foods such as soy sauce and fermented meat also generate kokumi substances resulting from protein breakdown (Kuroda and Miyamura 2015; Forde 2016; Zhao, Schieber, and Gänzle 2016). Therefore, adding kokumi compounds to beef broth or chicken broth results in better palatability as the meat flavor, odor, savory sensation, mouthfulness, thickness, and aftertaste sensory properties are enhanced. This enhancement of sensory qualities may also increase the satiety of the food, which may contribute to better satisfaction in low-calorie foods (Tang et al. 2020).

Considering the source of food and fermenting microbial cultures, several Asian seasonings, which are fermented products, are likely to contain kokumi substances. These seasonings are, for example, soy sauce, fish sauce, and fermented shrimp paste, which have already been mentioned as sources of umami substances. Fermented food such as fermented shrimp paste (from Indonesia, the Philippines, and China), various types of Japanese soy sauce, and fish sauce (from Vietnam, Thailand, China, Japan, and Italy) typically contain kokumi substances. The kokumi peptides are found in all fermented shrimp paste, varying from 0.9 to 5.2 µg/g. The fermentation of plant materials, like soy sauce, also contributes to kokumi peptides 0.15–0.61 mg/dL. However, fish sauce samples contain up to 1.26 mg/dL, but not all fish sauce samples contain kokumi peptides. Five brands of Thai fish sauce were tested; all were made from anchovies, and salt and kokumi peptides were detected in all samples (Kuroda et al. 2012, 2013; Miyamura et al. 2014). Kokumi peptides, in traditional northeastern-style freshwater fish (*pla-ra*, ปลาร้า), were identified and quantified in a total of ten commercial samples. A comparison of sensory analysis between the mixtures of *pla-ra* in water and in the model broth showed that an addition of *pla-ra* contributed to an increase in sensation and enhanced the salty and umami tastes (Phewpan et al. 2019). Common pla-ra recipe can be found in topics 7.5 and 7.11)

Seasoning derived from fermented fish products, especially *Nam-pla*, is evidently a crucial foundation of Thai cuisine. It provides saltiness, umami, kokumi, and a aromatic attributes that other products cannot replace. On the contrary, when the Thai dish is adjusted to fit people from the West's pallet, the fermented sauce is excluded. For example, a relatively newly invented menu item called "American fried rice" sported rice seasoned with salt and ketchup.

3.3 Aroma

Via the nose, receptors in our olfactory system are able to detect odorous volatile molecules even at very low concentrations, thus creating endless smell sensations. It is estimated that humans can distinguish more than 1 trillion different smells (Bushdid et al. 2014). However, out of more than 15,000 volatile compounds from foods that have been identified, only about 3,500 of those have detectable odorant notes (Guichard and Salles 2016). Nevertheless, a review of how much sense we detect from food confirms that the sense of smell is dominant (Spence 2015).

Recognition and identification of an individual odor can be limited, especially in a complex mixture of several volatile compounds. Rather, it is the integration of volatile compounds that creates a flavor impression. The flavor description tends to be for the entire food rather than specifying dominant odors from each ingredient (Lawless 1999). In the case of various Thai curries, for example, though there are many overlapping ingredients, most people will not recognize the existence of garlic or galangal but rather identify them as red curry or green curry.

While we can only taste food once it is in the mouth, we can perceive the smell of food before and during eating. Sniffing through the nose (the orthonasal route) sends the odor molecules higher into the nasal chamber to the receptors of the olfactory system. Once the food is being chewed in the mouth, the teeth break it down into smaller pieces, releasing more aromas and mixing it with the saliva. At the same time, the force needed to chew the food depends on the food texture and the feeling in the oral cavity, creating a sense of mouthfeel.

Another set of odor molecules travels through the back of an oral cavity in the mouth to the nasal chambers (the retronasal route). In other words, the inhaling path is orthonasal, while airflow from breathing out is retronasal. Through the retronasal route, we perceive most of the sensations from eating food or what we refer to as flavors of food (Lawless and Heymann 2010). Inhaling the smell of the food does not really convey what the food flavor will be like until a person chews the food and perceives the flavor through the retronasal route. Our brain cannot distinguish between the signals it receives from the taste receptors on the tongue and those signals from the smell receptors at the back of the nose, so the brain interprets both signals simply as taste from the mouth. Thus, the evaluation of food, even by a chef, generally refers to its "taste," while actually, it is the smell perceived while chewing the food that contributes to pleasure and thereby to the overall appreciation and liking.

To reach the nose cavity, the molecules have to be volatile. So the study of volatile compounds is the basis of understanding and identifying aroma substances or the molecules responsible for odors in a particular food or ingredient. Aroma substances presented in high enough concentration (more than the odor threshold at the lowest detectable level by a human) are likely to form the characteristic aroma of each food item or are considered the key odorant. Distinctive sources of aroma in Thai dishes are herbs and spices. These intense aromatic ingredients are presented as a mixture with different levels of complexity. More details on responsible compounds are given in Chapter 4, cooking techniques to enhance their strength are given in Chapter 5, and the roles of the aroma on the pleasantness of a Thai meal are given in Chapter 6.

3.4 Chemesthesis

Some chemicals can induce a sensation, such as a sense of touch and temperature, on our skin and evoke these senses in oral and nasal cavities. These chemicals are called chemesthesis. Many of the chemesthesis stimulate the trigeminal nerve in the mouth, nose, and eyes and send the signal of pain to the brain. So the sensations of burning, stinging, biting, numbing, and tingling are considered trigeminal senses. Examples are a hot-burning sensation from chili peppers, non-heat-related irritations from wasabi and mustard, tear-inducing irritation from onion, a cooking sensation from menthol, and fizziness from carbon dioxide in soda, beer, and sparkling wine. The sensation that symbolizes Thai foods is the pungency or the hot

Table 3.1A Photos, common name, Thai name, and scientific name of some common Thai herbs and spices

Common name
Bird's Eye Chili
Thai name
พริกขี้หนู (Prig-Kee-Noo)
Scientific name
Capsicum annuum

Common name
Spur Chili Pepper
Thai name
พริกขี้ฟ้า (Prig-Chee-Fah)
Scientific name
Capsicum annuum Linn. var acuminatum Fingerh.

Common name
Shallot
Thai name
หอมแดง (Hom-Dang)
Scientific name
Allium ascalonicum

Common name
Garlic
Thai name
กระเทียม (Kra-Tiam)
Scientific name
Allium sativum linn.

Table 3.1B

	Common name
	Holy Basil
	Thai name
	กะเพรา (Bai-Ka-Prao)
	Scientific name
	Ocimum sanctum

	Common name
	Sweet Basil
	Thai name
	โหระพา (Bai-Ho-Ra-Pa)
	Scientific name
	Ocimum basilicum

	Common name
	Kaffir Lime Rind
	Thai name
	ผิวมะกรูด (Pew-Ma-Krood)
	Scientific name
	Citrus hystrix

	Common name
	Lemongrass
	Thai name
	ตะไคร้ (Ta-Krai)
	Scientific name
	Cymbopogon citratus

burning sensation induced by chili. The other important class of chemesthesis associated with Thai food is astringency. Chemicals responsible for astringency are tannins and polyphenolic compounds in various herbs and spices. See list of common herbs and spices in Thai cuisine in Tables 3.1A-D. The astringency is considered a tactile sensation because it is the sense of touch that we feel in the mouth. The sensation can be described as feeling rough and dry in the mouth with puckering, tightening feel on the tongue and cheeks (Lawless and Heymann 2010; McDonald, Bolliet, and Hayes 2016).

Table 3.1C

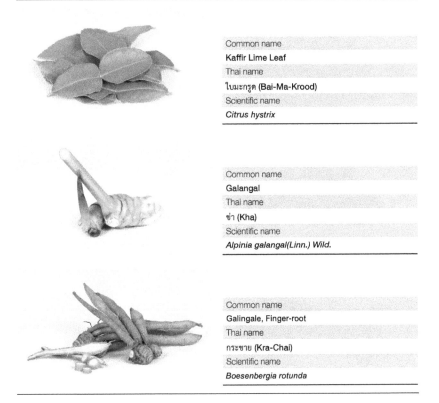

Common name	
Kaffir Lime Leaf	
Thai name	
ใบมะกรูด (Bai-Ma-Krood)	
Scientific name	
Citrus hystrix	
Common name	
Galangal	
Thai name	
ข่า (Kha)	
Scientific name	
Alpinia galangal(Linn.) Wild.	
Common name	
Galingale, Finger-root	
Thai name	
กระชาย (Kra-Chai)	
Scientific name	
Boesenbergia rotunda	

3.4.1 Pungency and Burning, the Chemical Heat

Though the sensation of heat and burning from chili is scientifically and literally considered a sensation of "pain," the sensation is desirable in several traditional cuisines, especially Thailand's. It is possible that Thai people, in particular, love hot and spicy food and form a

Table 3.1D

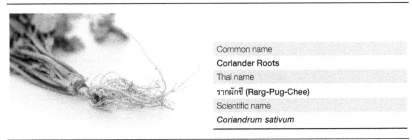

Common name	
Coriander Roots	
Thai name	
รากผักชี (Rarg-Pug-Chee)	
Scientific name	
Coriandrum sativum	

Table 3.1E

	Common name
	Cardamom
	Thai name
	ลูกกระวาน (Loog-Kra-Wan)
	Scientific name
	Amomum testaceum Ridl.

	Common name
	Cloves
	Thai name
	กานพลู (Karn-Plu)
	Scientific name
	Syzygium aromaticum

	Common name
	Cinnamon
	Thai name
	อบเชย (Ob-Choey)
	Scientific name
	Cinnamomum Spp.

	Common name
	Star Anise
	Thai name
	โป๊ยกั๊ก (Poy-Kug)
	Scientific name
	Illicium verum

familiarity or preference for the flavor even from the time they were in their mother's womb.

The development process at NASA requires a panel of astronauts to rate the menu using a 9-point hedonic scale, i.e., a score from 1, meaning dislike extremely, to a score of 9, for like extremely. Only a score of 6 will be developed further and be constantly taste-tested before final selection.

The hot and spicy character is desirable for those who are not familiar with Thai food. Rachael Ray, a world-renowned chef, was

Table 3.1F

Common name	
Cumin	
Thai name	
ยี่หร่า (Yee-Rah)	
Scientific name	
Cuminum cyminum	

Common name	
Nutmeg	
Thai name	
ลูกจัน (Loog-Chun)	
Scientific name	
Myristica fragrans	

Common name	
Mace	
Thai name	
ดอกจัน (Dok-Chun)	
Scientific name	
Myristica fragrans	

Table 3.1G

Common name	
Pepper	
Thai name	
พริกไทยดำ (Prig-Tai)	
Scientific name	
Piper nigrum L.	

Common name	
Coriander seeds	
Thai name	
ลูกผักชี (Loog-Puk-Chee)	
Scientific name	
Coriandrum sativum	

Table 3.1H

Common name	Bay Leaf
Thai name	ใบกระวาน (Bai-Kra-Waan)
Scientific name	*Cinnamomum porectum*

invited to NASA to develop space food. One of her menus, "Rachael Ray's spicy Thai chicken with red peppers and basil," was designed for astronauts to eat in space. The recipe is listed as the main dish in *The Astronaut's Cookbook* (Bourland and Vogt 2009). The dish has to be freeze-dried in order to be packaged for spaceflight. The recipe also contains herbs and spices such as garlic, chili, pepper, and basil, just like one of the most popular hot and spicy Thai dishes, "chicken basil and rice" (*khao-kra-prao-gai*, ข้าวกระเพราไก่).

The NASA-Advanced Food Technology Project has been working on supplying food for a mission to Mars, a journey to the red planet in the 2030s with six to eight astronauts spending almost three years in space. The NASA food scientists have to make sure that food not only has at least five years of shelf life. Also, they have to make sure that, psychologically, the food is appealing enough for astronauts to eat and gain sufficient nutrients so they can stay healthy and happy throughout the years in space (Cooper, Douglas, and Perchonok 2011). Of more than 100 vegetarian recipes tested, the most popular one among the astronaut panel is "Thai Pizza." The dish has various vegetables plus red pepper and spicy sauce (BBC Newsround 2012; Plushnick-Masti 2012; Spector 2012). Though the dish under development is not authentic Thai cuisine, its popularity suggests that spiciness is a characteristic of Thai food. Furthermore, chili is one of the essential ingredients associated with Thai cuisine.

Information on the Thai pizza being developed in NASA's lab made big headlines in 2012. Then, in 2019, with the attempt to grow fresh crops on Mars, chili peppers (*Capsicum annuum*) were selected. The selection made chili peppers the first fruit (after other vegetables in the NASA vegetable-production system) to grow on another planet.

In addition to the botanical possibilities, another important reason for choosing chilies is to prevent food fatigue. Astronauts will have limited menu varieties for those years, and also, in space, their sensitivity for detecting flavors and enjoying food will tend to be lower than on earth. So, foods tend to taste bland due to the microgravity environment. Therefore, more spicy foods, with a hot flavor and more flavor overall, are desirable for astronauts (Bowler 2019; Mervosh 2019).

Another profound Thai flavor concept characterized by the hot and burning sensation of chili is the Sriracha chili sauce. The hot and spicy sauce obtained from the fermentation of chili, pickled garlic, salt, and sugar was originally developed in the Si Racha District of Chonburi Province. This seaside town with fishing villages has many kinds of seafood, and the chili sauce was explicitly developed to accompany seafood dishes. The chili sauce production was mainly on a small household scale. In 1932, Gold Medal Brand Sriracha Sauce (ตราเหรียญทอง) launched commercially and had since claimed to be "the first in Siam" to develop the Sriracha sauce (Sundaravej 2021). Another brand, Sriraja Panich (ศรีราชาพานิช), was launched around the same time but has a much larger market share since its merge with a large sauce-manufacturing company (Sriraja Panich 2021). Both the Sriraja Panich and the Gold Medal brands declared similar ingredients, which are red chili, pickled garlic (*gra-tium-dong*, กระเทียมดอง), sugar, vinegar, and salt. In addition, both manufacturing processes required fermentation of the ingredients. There is no starch or any chemical additives such as preservative, flavor enhancer, or colorant.

Fifty years after the commercial launch of Sriracha sauce, other brands of Sriracha chili sauce were manufactured outside of Thailand. These chili sauces have bright orangey-red color, a signature of the Sriracha sauce. They also contain similar key ingredients as the original Sriracha sauce. However, garlic was used instead of garlic pickle, and it is unknown whether the product is fermented. In addition, certain brands may contain chemical food additives such as potassium sorbate, sodium bisulfite, and xanthan gum. According to the Sriracha Panich website, the chilies used in the processing is long red pepper or spur chili (*prig-chee-fah*, พริกชี้ฟ้า) (Sriraja Panich 2021), unlike the Sriracha sauce abroad, which used the jalapenos peppers (Hammond 2013).

Though formulated initially to complement seafood, the Sriracha dipping sauce is now used with several food items. Dishes that need the Sriracha sauce are, for example, savory omelet (*kai-jeow*, ไข่เจียว), stir-fried noodles, and chicken- (*guay-teow-kua-gai*, ก๋วยเตี๋ยวคั่วไก่) and oyster-crispy omelet (*hoy-tod*, หอยทอด).

"Honey Sriracha" flavor was on the list of the fastest-growing flavors in restaurants in the What's Hot 2019 Culinary Forecast by the National Restaurant Association (Sloan 2019). This forecast should not come as a surprise since, back in 2015, Money magazine had an article entitled "Why you should blame millennials for spicy fast food." The report listed fast-food chains in the United States such as Taco Bell, Pizza Hut, and Denny's as those incorporating Sriracha sauce in their menus. The action was considered to be quite late, considering that McDonald's, Sonic, Burger King, Wendy's, and Subway had already done so since 2012 (Tuttle 2015).

Also, starting in 2012, Frito Lay launched a campaign called the "Do Us A Flavor" contest for the best-suggested flavor of Lay's potato chips. According to the Frito Lay website, millions of chip flavor ideas were received. Sriracha flavor was one of the top three finalists voted in the United States, surpassing those other millions of ideas. As a result, Sriracha Lay's was developed into a commercial product and launched to the public in 2012 (Frito Lay 2015). Seven years later, in 2018, Frito Lay launched another Sriracha flavored product in Japan, Sriracha Doritos, and then, in 2019, the Sriracha Doritos was also launched in the United States (Bjornson 2019). Several factors had anticipated the craze in Sriracha. As pointed out in the Money magazine article, the change was probably led by millennials who are likely to see themselves as "adventurous eaters" and consider Sriracha sauce a "go-to condiment." The intrinsic factor probably is the flavor profile of the sauce itself, which could be described as Thai hot sauce. Still, with sweet heat, *"similar to the allure of bacon, sriracha is now being used as an unorthodox flavor enhancer for everything from snack food to candy and alcohol."* (Hensel 2014).

Chili contains capsaicinoids that are mainly capsaicin and dihydrocapsaicin. The chemicals are considered oral irritants, not flavor compounds. Upon consumption, the burning sensation produced in the oral cavity evokes other responses, especially the body's defensive reaction with teary eyes, runny nose, face turning red, and sweating. The sensation, which may last for several minutes, can be described as warming, burning, stinging, and tingling on the tongue and in the oral cavity and throat. The experience should dissuade humans from eating the chili. But after the initial exposure, we realize that, ultimately, there is no real harm. Eating food with pungency from chili is compatible with riding roller coasters. The joy comes from the thrill we receive from taking "constrained risks." Increased exposure to chili, especially when associated with positive outcomes such as improving the taste of bland foods or the positive experience of dining socially, can gradually turn those hostile to chili into likers. The perceived burning sensation from capsaicin is higher among people who rarely consume chili pepper than those who regularly eat chili. However, liking or disliking chili tends to be influenced by early exposure rather than physiological adaptation (Rozin and Schiller 1980; Lawless, Rozin, and Shenker 1985; Reinbach, Toft, and Møller 2009; Ludy and Mattes 2012; Spence 2018).

In food cultures with chili consumption habits such as Thailand, exposure to the pungency of the chili may even start before birth. Other substances that create a hot sensation are, for example, piperine from black pepper and zingerone from ginger. Both pepper and ginger are also common ingredients of Thai food. The preference for certain foods or flavors may have been "programmed" in the fetus. The exposure to the flavors is from the amniotic fluid during pregnancy and from breast milk during the lactation period (Mennella, Jagnow, and Beauchamp 2001; Mennella 2009). In order to turn from disliking to liking certain foods, it may take up to 10–20 exposures. The number drops if the first exposure is at an early age (Cooke and Fildes 2011). In this regard, most Thai people may have already enjoyed these flavors since the time they were in their mother's wombs. Those who are non-eaters of chili can later learn to like the pungency along with other flavors in the culture, but repeated exposure may be needed.

Prolonged exposure to hot and spicy food may affect how Thai people taste food. There are many findings on reducing sensitivity to flavor stimuli, though all the results are not consistent (Lawless, Rozin, and Shenker 1985; Carstens et al. 2007). A more recent study was conducted with the observation of significant differences between Thai and Japanese cuisines. Most Thai dishes are a mixture of several flavors from many ingredients, usually including chili for heat and spiciness. The Japanese, however, like to separate dishes and prefer only one prominent flavor per dish, with no chili or hot sauce. Most Thai participants in the study liked spicy food and ate spicy food at least weekly, while the Japanese rarely did. Thus, it was no surprise to see that Thai participants needed higher levels of sweet, salty, sour, bitter, and umami tastes than the Japanese did to be able to detect those tastes (Trachootham et al. 2017). There is not enough data to draw a firm conclusion, but it is likely that frequent consumption of hot and spicy food decreases flavor detection sensitivity.

3.4.2 Astringency

Substances that can stimulate an astringency sensation (*fhard*, ฝาด) are mostly the polyphenolic compounds found in plants, especially herbs and spices. There are more than 8,000 polyphenolic compounds in the human diet that have already been identified. This large class of substances refers to molecules with phenolic structure (organic aromatic compounds), which convey their ability to act as antioxidants (Tsao 2010). The polyphenol content can be used to predict the astringency of the food samples. The rate of astringency can be predicted by polyphenol-protein complex, using mucin, the slippery compound found in saliva, as a protein source (Monteleone et al. 2004).

Coffee and tea are rich in polyphenols and these beverages, when consumed without milk or sugar, are bitter and astringent. Unlike these popular beverages, the sensation of astringency from plant-based foods such as vegetables, herbs, and spices is not always pleasant nor desirable. As the polyphenolic compounds, especially tannins, bind with proteins and mucin in the saliva, the oral cavity loses its slippery coating, thereby causing the lips and mouth to feel dry, the oral tissues to feel rough and puckered, and even creating a pulling sensation in the muscles of the cheeks. However, the not-so-pleasant sensation is a trade-off with the many health benefits of those phytochemicals in plant-based food (Lee and Lawless 1991; Byrne 2016).

The consumption of polyphenols from fruit, vegetables, and whole grains (phytochemicals) is encouraged since it has been proven to lower the risk of degenerative diseases, especially those caused by oxidation. For example, cancer, cardiovascular disease, and diabetes inhibit histamine release, and some also have antibacterial and antiviral properties. One serving of Thai food and beverages, such as a cup of ginger tea, one serving of yellow curry, has as much antioxidant activity as a serving of fruits and vegetables (Sinsawasdi 1998). In addition, the combination of polyphenol antioxidants tends to have a synergistic effect. Combining varieties of plant-based foods displays better anticancer activity, and the result is better than those antioxidants taken from food supplement pills (Kaefer and Milner 2008; Shahidi and Ambigaipalan 2015). Consistent with other reports on the health benefits of polyphenols, edible parts of a plant consumed have been studied for antioxidant activity, flavonoid content, and antimutagenicity (Murakami, Ohigashi, and Koshimizu 1994; Nakahara et al. 2002; Maisuthisakul, Pasuk, and Ritthiruangdej 2008; Sriket 2014). Apart from fruit and vegetables, varieties of herbs and spices used in Thai food, such as lemongrass, ginger, cilantro, and chili peppers, contain high polyphenols effective as antioxidants and preservatives (Kaefer and Milner 2008; Nakornriab and Puangpronpitag 2011; Harmayani et al. 2019).

Although Thai cuisine usually contains ingredients with a wide variety and quantity of polyphenolic compounds, the astringency and, in some dishes, bitterness induced by these compounds do not always result in less liking. In fact, it is the characteristic expected in a combination of taste and smell, along with the texture and color of each recipe (Kanchanakunjara et al. 2017). Just like other pungent, astringent, or bitter food and drinks (like beer, wine, tea, or coffee), preference for them can be changed. The initial response may be negative, but the affection for the food can be increased with more exposure. The food exposure tactic to increase liking is effective in both children and adults. Other factors such as health benefits, advertising, and social context can also increase preference (Lesschaeve and Noble 2005).

3.5 Interactions of Senses and Flavor Perception

Just about everyone has some common experience of sensory interaction. A smell can enhance certain tastes. Sweeter odors can give a perception of sweeter taste. A better visual presentation of food provides us with an expectation that food is more flavorful. Adding artificial colorant to food and drink can lead to a higher expectation of more intense taste and flavor.

In an experiment on the lowest concentration of detectable taste (threshold concentration), the threshold for the sweetness and saltiness (a solution of sucrose, sodium chloride) was unaffected by either MSG or IMP. The threshold for sourness (a solution of tartaric acid) was increased by the addition of MSG and IMP. The minimum detection level of bitterness and umami (a solution of quinine sulfate and MSG, respectively) was lower with the addition of IMP (Yamaguchi and Ninomiya 2000). MSG was found to increase the threshold for sourness but did not change those of sweetness, saltiness, and bitterness. IMP increased the minimum detectable concentration of sourness and bitterness, but the sweetness and saltiness thresholds were not affected. Thus, it is likely that the addition of glutamate can lower our sensitivity to perceiving sourness in food; in other words, food tastes less sour. When nucleotides are present, the sourness and bitterness are reduced, while the umami taste is intensified. The synergistic effect of glutamate and nucleotides has been studied and confirmed in many experiments.

The reduction of salt is a challenge. Not only that it decreases saltiness perception but it also reduces the overall acceptance and intensity of the flavor of the food. Umami can enhance saltiness in food. Thus, the addition of MSG helps to reduce overall sodium intake in various foods without compromising palatability.

The aroma can also interact with taste and enhance the taste perception if the flavor is salt-congruent, so the interaction depends on how well the smell goes with the food. For example, in a commercial bouillon cube for meat broth preparation, the addition of a savory aroma that gave a brothy, meaty, or roasted note was salt-congruent. By up to 30%, reducing salt from the chicken or beef bouillon cube without changing its flavor profile is possible with the addition of flavor that characterizes the broth (Lawrence et al. 2009; Batenburg and van der Velden 2011).

One ingredient that is essential to Thai food is fish sauce. Though its primary function is to add saltiness to various dishes, it also provides a specific aroma to food and adds several dimensions of sensory qualities, especially umami and kokumi. It is very common for Thai people to carry a small bottle or packet of fish sauce when traveling overseas. The palatability of food and, in some dishes, the identity of that particular dish is lost if fish sauce is replaced with salt or other seasonings.

The aromatic compounds in herbs and spices are more hydrophobic, so their molecules tend to exist in fatty components of the food. These volatile compounds are released in the air phase during chewing in the mouth, eventually, rise through the retronasal route, and are perceived by the olfactory receptors in the nose. Fats with shorter chain fatty acids tend to retain these volatile compounds in food better, and the release of the volatiles to the nose is better if the food has no emulsifier (Guichard and Salles 2016). Vegetable oils have more long-chain fatty acid, and processed coconut milk has added emulsifiers. On the other hand, freshly squeezed coconut cream contains short-chain fatty acids with no emulsifier. Hence, fresh coconut cream tends to be a more favorable ingredient for extracting and delivering food aromas. In other words, dishes like curry tend to have higher perceivable flavor intensity when cooked with fresh coconut cream.

With the vast variety of food experiences each person is exposed to, identifying and recognizing any food requires an association between flavor and memory recall. The process of perceiving, recognizing, and liking of flavor is further complicated by various factors, such as a person's mood during specific food consumption, prenatal exposure to certain foods, and frequency of consumption. The multisensory process is so complicated that even recognizing one's own ethnic food might not always be accurate. For example, in an experiment where American participants (in the United States) and Thai participants (in Thailand) were asked to identify various spice blends, only 53% of the participants were able to identify the Thai *tom-yum* sample correctly (Bell et al. 2011). Though this was slightly better than the American counterparts (19% correct), the evidence shows that it takes more than taste and smell to accurately identify any food item.

Eating is one of the most enjoyable and pleasurable activities for humans. The interactions of taste and odor, irritation and flavor, and color and flavor have been well documented scientifically. Multisensory integration impacts before and during the consumption experience and greatly influences the personal liking of food. The cycle is composed of wanting to eat (hunger), judging the food being eaten (liking), and feeling experience after eating (learning, satisfying). People usually evaluate food by its appearance, orthonasal smell, texture in the mouth while masticated, and flavor. Eating is a learning experience that potentially changes a person's food preference if the learning process is pleasurable (Fjældstad, van Hartevelt, and Kringelbach 2016). Elements in the Thai meal experience will be further discussed in detail in Chapter 6 as the pleasure from eating food is stimulated not only by certain flavors but also by the total experience of all senses.

3.6 Conclusion

The deliciousness, especially from umami taste, is the foundation of many Thai dishes. Therefore, products of fish fermentation used as seasoning are key ingredients that cannot be replaced with salt. Most foods stimulate two primary chemical senses, i.e., gustatory (taste) and olfactory (smell). However, Thai food evokes an additional trigeminal sensations with its relatively high chemesthesis content. The efficiency of trigeminal system stimulation lies not only in the herb and spice mixture but also in the delivery system, such as the use of coconut milk in many recipes. The most remarkable ingredient is chili pepper as its spiciness is almost synonymous with Thai cuisine. Evidence pointed out that local people may have formed a tendency toward spicy food liking from before birth. For foreigners, it may take several exposures before one can learn to like the burning sensation of Thai food.

References

Batenburg, Max, and Rob van der Velden. 2011. "Saltiness enhancement by savory aroma compounds." *Journal of Food Science* 76 (5):S280–S288.

BBC Newsround. 2012. "What's on the Menu for a Mission to Mars?" https://www.bbc.co.uk/newsround/18887663

Bell, Brandon, Koushik Adhikari, Edgar Chambers, Panat Cherdchu, and Thongchai Suwonsichon. 2011. "Ethnic food awareness and perceptions of consumers in Thailand and the United States." *Nutrition & Food Science* 4 (41): 268–277.

Bjornson, Greta. 2019. "Sriracha Doritos Have Arrived for a Limited Time." *People*, accessed January 8. https://people.com/food/sriracha-doritos/.

Bongsanid, Yaovabha. 1935. *Tumrub Sai Yaovabha (ตำรับสายเยาวภา)*. Bangkok, Thailand: Saipunya Sakakom.

Bourland, Charles T, and Gregory L Vogt. 2009. "Main dishes." In *The Astronaut's Cookbook*, 82–116. New York, NY: Springer.

Bowler, Jacinta. 2019. "NASA Has Announced the First Fruit They'll Grow on the ISS, and It's Hot." *ScienceAlert*. https://www.sciencealert.com/nasa-has-announced-the-first-fruit-they-want-to-grow-on-the-iss.

Bushdid, Caroline, Marcelo O Magnasco, Leslie B Vosshall, and Andreas Keller. 2014. "Humans can discriminate more than 1 trillion olfactory stimuli." *Science* 343 (6177):1370–1372.

Byrne, Brian. 2016. "Interactions in chemesthesis: Everything affects everything else." In *Chemesthesis: Chemical Touch in Food and Eating*, edited by Shane Thomas McDonald, David Bolliet, John E. Hayes, 154–165. Chichester, West Sussex, England: Wiley Blackwell.

Carstens, Earl, Kelly C Albin, Christopher T Simons, and Mirela Iodi Carstens. 2007. "Time course of self-desensitization of oral irritation by nicotine and capsaicin." *Chemical Senses* 32 (9):811–816.

Chi, SP, and TC Chen. 1992. "Predicting optimum monosodium glutamate and sodium chloride concentrations in chicken broth as affected by spice addition." *Journal of Food Processing and Preservation* 16 (5):313–326. doi: 10.1111/j.1745-4549.1992.tb00212.x

Chotechuang, Nattida, Matichon Lokkhumlue, and Pawinee Deetae. 2018. "Effect of temperature and time on free amino acid profile in Thai chicken bone soup stock preparation." *Thai Journal of Pharmaceutical Sciences (TJPS)* 42 (3): 110–117.

Cooke, L, and A Fildes. 2011. "The impact of flavour exposure in utero and during milk feeding on food acceptance at weaning and beyond." *Appetite* 57 (3):808–811. doi: doi.org/10.1016/j.appet.2011.05.317

Cooper, Maya, Grace Douglas, and Michele Perchonok. 2011. "Developing the NASA food system for long-duration missions." *Journal of Food Science* 76 (2):R40–R48. doi: 10.1111/j.1750-3841.2010.01982.x

Curtis, Robert I. 2009. "Umami and the foods of classical antiquity." *The American Journal of Clinical Nutrition* 90 (3):712S–718S. doi: 10.3945/ajcn.2009.27462C

Fjældstad, Alexander, Tim J van Hartevelt, and Morten L Kringelbach. 2016. "Pleasure of food in the brain." In *Multisensory Flavor Perception*, edited by Piqueras-Fiszman Betina, Spence Charles, 211–234. New York, NY: Elsevier.

Forde, Ciarán G. 2016. "11 – Flavor perception and satiation." In *Flavor*, edited by Patrick Etiévant, Elisabeth Guichard, Christian Salles and Andrée Voilley, 251–276. Elsevier, Cambridge, USA: Woodhead Publishing.

Frito Lay. 2015. "Lay's Brand Announces Last Call To Submit Next Great Potato Chip Flavor Idea For Chance At $1 Million Grand Prize." Accessed January 8. https://www.fritolay.com/news/lay-s-brand-announces-last-call-to-submit-next-great-potato-chip-flavor-idea-for-chance-at-1-million-grand-prize.

Giacometti, T., L. J. Filer, S. Garattini, and M. R. Kare. 1979. "Free and bound glutamate in natu ral products." In *Glutamic Acid: Advances in Biochemistry and Physiology*, edited by J Filer Lloyd, New York: Raven Press, 25–34.

Guichard, Elisabeth, and Christian Salles. 2016. "Retention and release of taste and aroma compounds from the food matrix during mastication and ingestion." In *Flavor*, edited by Patrick Etiévant, Elisabeth Guichard, Christian Salles, Andrée Voilley, 3–22. Amsterdam: Elsevier.

Hajeb, P, and S Jinap. 2015. "Umami taste components and their sources in Asian foods." *Critical Reviews in Food Science and Nutrition* 55 (6):778–791. doi: 10.1080/10408398.2012.678422

Hammond, Griffin. 2013. "Sriracha, A Documentary Film by Griffin Hammond." tumblr.

Harmayani, Eni, Anil Kumar Anal, Santad Wichienchot, Rajeev Bhat, Murdijati Gardjito, Umar Santoso, Sunisa Siripongvutikorn, Jindaporn Puripaatanavong, and Unnikrishnan Payyappallimana. 2019. "Healthy food traditions of Asia: Exploratory case studies from Indonesia, Thailand, Malaysia, and Nepal." *Journal of Ethnic Foods* 6 (1):1. doi: 10.1186/s42779-019-0002-x

Hartley, Isabella E, Djin Gie Liem, and Russell Keast. 2019. "Umami as an 'alimentary' taste. A new perspective on taste classification." *Nutrients* 11 (1):182.

Hensel, Kelly. 2014. "Top 5 flavor trends." *Food Technology*, November 1, 68: 11.

Horio, Tsuyoshi, and Yojiro Kawamura. 1990. "Studies on after-taste of various taste stimuli in humans." *Chemical Senses* 15 (3):271–280.

Huynh, Hue Linh, Robert Danhi, and See Wan Yan. 2016. "Using fish sauce as a substitute for sodium chloride in culinary sauces and effects on sensory properties." *Journal of Food Science* 81 (1):S150–S155.

Jang, SooCheong Shawn, Aejin Ha, and Carol A Silkes. 2009. "Perceived attributes of Asian foods: From the perspective of the American customers." *International Journal of Hospitality Management* 28 (1):63–70.

Kaefer, Christine M, and John A Milner. 2008. "The role of herbs and spices in cancer prevention." *The Journal of Nutritional Biochemistry* 19 (6):347–361. doi: https://doi.org/10.1016/j.jnutbio.2007.11.003

Kanchanakunjara, Taddara, Chantachon, Songkoon, Koseyayotin, Marisa. 2017. "The evolution of Thai curry pastes." *Dusit Thani College Journal* 11 (special):249–266.

Keast, Russell SJ, and Andrew Costanzo. 2015. "Is fat the sixth taste primary? Evidence and implications." *Flavour* 4 (1):5.

Kuninaka, A. 1960. "Studies on taste of ribonucleic acid derivatives." *Journal of the Agricultural Chemical Society of Japan* 34:489–492.

Kure Liu, Christopher, Paule Valery Joseph, Dana E Feldman, Danielle S Kroll, Jamie A Burns, Peter Manza, Nora D Volkow, and Gene-Jack Wang. 2019. "Brain imaging of taste perception in obesity: A review." *Current Nutrition Reports* 8 (2):108–119. doi: 10.1007/s13668-019-0269-y

Kurihara, Kenzo. 2009. "Glutamate: From discovery as a food flavor to role as a basic taste (umami)." *The American Journal of Clinical Nutrition* 90 (3):719S–722S. doi: 10.3945/ajcn.2009.27462D

Kuroda, Motonaka, Yumiko Kato, Junko Yamazaki, Yuko Kai, Toshimi Mizukoshi, Hiroshi Miyano, and Yuzuru Eto. 2012. "Determination and quantification of γ-glutamyl-valyl-glycine in commercial fish sauces." *Journal of Agricultural and Food Chemistry* 60 (29):7291–7296.

Kuroda, Motonaka, Yumiko Kato, Junko Yamazaki, Yuko Kai, Toshimi Mizukoshi, Hiroshi Miyano, and Yuzuru Eto. 2013. "Determination and quantification of the kokumi peptide, γ-glutamyl-valyl-glycine, in commercial soy sauces." *Food Chemistry* 141 (2):823–828.

Kuroda, Motonaka, and Naohiro Miyamura. 2015. "Mechanism of the perception of "kokumi" substances and the sensory characteristics of the "kokumi" peptide, γ-Glu-Val-Gly." *Flavour* 4 (1):11.

Lawless, Harry, Paul Rozin, and Joel Shenker. 1985. "Effects of oral capsaicin on gustatory, olfactory and irritant sensations and flavor identification in humans who regularly or rarely consume chili pepper." *Chemical Senses* 10 (4):579–589. doi: 10.1093/chemse/10.4.579

Lawless, Harry T. 1999. "Descriptive analysis of complex odors: Reality, model or illusion?" *Food Quality and Preference* 10 (4–5):325–332.

Lawless, Harry T, and Hildegarde Heymann. 2010. "Physiological and psychological foundations of sensory function." In *Sensory Evaluation of Food*, edited by T. Lawless Harry, Heymann Hildegarde, 19–56. New York, NY: Springer.

Lawrence, G, C Salles, C Septier, J Busch, and T Thomas-Danguin. 2009. "Odour–taste interactions: A way to enhance saltiness in low-salt content solutions." *Food Quality and Preference* 20 (3):241–248. doi: 10.1016/j.foodqual.2008.10.004

Lee, Christopher B, and Harry T Lawless. 1991. "Time-course of astringent sensations." *Chemical Senses* 16 (3):225–238. doi: 10.1093/chemse/16.3.225

Leong, Jasmine, Chinatsu Kasamatsu, Evelyn Ong, Jia Tse Hoi, and Mann Na Loong. 2016. "A study on sensory properties of sodium reduction and replacement in Asian food using difference-from–control test." *Food Science & Nutrition* 4 (3):469–478.

Lesschaeve, Isabelle, and Ann C Noble. 2005. "Polyphenols: Factors influencing their sensory properties and their effects on food and beverage preferences." *The American Journal of Clinical Nutrition* 81 (1):330S–335S.

Ludy, Mary-Jon, and Richard D Mattes. 2012. "Comparison of sensory, physiological, personality, and cultural attributes in regular spicy food users and non-users." *Appetite* 58 (1):19–27.

Maisuthisakul, Pitchaon, Sirikarn Pasuk, and Pitiporn Ritthiruangdej. 2008. "Relationship between antioxidant properties and chemical composition of some Thai plants." *Journal of Food Composition and Analysis* 21 (3):229–240. doi: https://doi.org/10.1016/j.jfca.2007.11.005

Marcus, Jacqueline B. 2009. "Unleashing the power of umami." *Food Technology* 63: 11.

McBurney, Donald H, and Janneane F Gent. 1979. "On the nature of taste qualities." *Psychological Bulletin* 86 (1):151.

McDonald, Shane T, David A Bolliet, and John E Hayes. 2016. "Introduction: What is chemesthesis?" *Chemesthesis*, 1–3. Chichester, West Sussex, England: Wiley Blackwell.

Meiselman, Herbert L. 2009. "Dimensions of the meal: A summary." In *Meals in Science and Practice*, 3–15. Woodhead Publishing.

Mennella, Julie A. 2009. "Flavour programming during breast-feeding." In *Breast-Feeding: Early Influences on Later Health*, 113–120. Springer.

Mennella, Julie A, Coren P Jagnow, and Gary K Beauchamp. 2001. "Prenatal and postnatal flavor learning by human infants." *Pediatrics* 107 (6):e88–e88.

Mervosh, Sarah. 2019. "A Menu for Mars? NASA Plans to Grow Chiles in Space." *The New York Times*. https://www.nytimes.com/2019/07/20/science/nasa-food-gardening-mars.html.

Miyamura, Naohiro, Motonaka Kuroda, Yumiko Kato, Junko Yamazaki, Toshimi Mizukoshi, Hiroshi Miyano, and Yuzuru Eto. 2014. "Determination and quantification of a kokumi peptide, γ-glutamyl-valyl-glycine, in fermented shrimp paste condiments." *Food Science and Technology Research* 20 (3):699–703.

Monteleone, Erminio, Nicola Condelli, Caterina Dinnella, and Mario Bertuccioli. 2004. "Prediction of perceived astringency induced by phenolic compounds." *Food Quality and Preference* 15 (7):761–769. doi: https://doi.org/10.1016/j.foodqual.2004.06.002.

Murakami, A, H Ohigashi, and K Koshimizu. 1994. "Possible anti-tumour promoting properties of traditional Thai food items and some of their active constituents." *Asia Pacific Journal of Clinical Nutrition* 3 (4):185–91.

Mustafa, Zannara, Sana Ashraf, Syeda Fareeha Tauheed, and Sikandar Ali. 2017. "Monosodium glutamate, commercial production, positive and negative effects on human body and remedies—A review." *International Journal of Scientific Research in Science and Technology* 3:425–435.

Nakahara, Kazuhiko, Gassinee Trakoontivakorn, Najeeb S Alzoreky, Hiroshi Ono, Mayumi Onishi-Kameyama, and Mitsuru Yoshida. 2002. "Antimutagenicity of some edible Thai plants, and a bioactive carbazole alkaloid, mahanine, isolated from *Micromelum minutum*." *Journal of Agricultural and Food Chemistry* 50 (17):4796–4802. doi: 10.1021/jf025564w

Nakornriab, Muntana, and Darunee Puangpronpitag. 2011. "Antioxidant activities and total phenolic contents of Thai curry pastes." *International Journal of Applied Chemistry* 7 (1):43–52.

Ninomiya, Kumiko. 1998. "Natural occurrence." *Food Reviews International* 14 (2–3):177–211.

Ninomiya, Kumiko, Shinichi Kitamura, Ai Saiga-Egusa, Shinichi Ozawa, Yuko Hirose, Tomoko Kagemori, Akira Moriki, Toshikazu Tanaka, and Toshihide Nishimura. 2010. "Changes in free amino acids during heating bouillon prepared at different temperatures." *Journal of Home Economics of Japan* 61 (12):765–773.

Pérez-Palacios, Trinidad, Joana Eusebio, Silvina Ferro Palma, Maria João Carvalho, Jorge Mir-Bel, and Teresa Antequera. 2017. "Taste compounds and consumer acceptance of chicken soups as affected by cooking conditions." *International Journal of Food Properties* 20 (sup1):S154–S165.

Phewpan, Apiniharn, Preecha Phuwaprisirisan, Hajime Takahashi, Chihiro Ohshima, Panita Ngamchuachit, Punnida Techaruvichit, Sebastian Dirndorfer, Corinna Dawid, Thomas Hofmann, and Suwimon Keeratipibul. 2019. "Investigation of Kokumi substances and bacteria in Thai fermented freshwater fish (Pla-ra)." *Journal of Agricultural and Food Chemistry* 68 (38): 10345–10351.

Plushnick-Masti, Ramit. 2012. "NASA Builds Menu for Planned Mars Mission in 2030s." Associated Press. http://www.nbcnews.com/id/48210630/ns/technology_and_science-space/t/nasa-builds-menu-planned-mars-mission-s/#.XhJ8dkeME2w.

Ratirita. 2018. ""Yum Yum" increased market share by reformulate tom-yum-koong flavored instant noodle." Brand Inside. https://brandinside.asia/yumyum-refresh-tom-yum-kung/.

Reinbach, Helene Christine, M Toft, and Per Møller. 2009. "Relationship between oral burn and temperature in chili spiced pork patties evaluated by time–intensity." *Food Quality and Preference* 20 (1):42–49.

Ritthiruangdej, Pitiporn, and Thongchai Suwonsichon. 2007. "Relationships between NIR spectra and sensory attributes of Thai commercial fish sauces." *Analytical Sciences* 23 (7):809–814.

Rogers, Peter J, and John E Blundell. 1990. "Umami and appetite: Effects of monosodium glutamate on hunger and food intake in human subjects." *Physiology & Behavior* 48 (6):801–804.

Rozin, Paul, and Deborah Schiller. 1980. "The nature and acquisition of a preference for chili pepper by humans." *Motivation and Emotion* 4 (1):77–101.

Ruddle, Kenneth, and Naomichi Ishige. 2010. "On the origins, diffusion and cultural context of fermented fish products in Southeast Asia." In *Globalization, Food and Social Identities in the Asia Pacific Region*, edited by James Farrer, 1–17.

Running, Cordelia A, Bruce A Craig, and Richard D Mattes. 2015. "Oleogustus: The unique taste of fat." *Chemical Senses* 40 (7):507–516.

Running, Cordelia A, and Richard D Mattes. 2016. "A review of the evidence supporting the taste of non-esterified fatty acids in humans." *Journal of the American Oil Chemists' Society* 93 (10):1325–1336. doi: 10.1007/s11746-016-2885-7

Sanitwong, Mom, and Ractchawongse Tuang. 1980. "The Important Things to Know (สิ่งสำคัญที่ควรทราบ)." In *Tumrub Sai Yaowapa (ตำรับสายเยาวภา)*, edited by Yaovabha Bongsanid, 34–35. Bangkok, Thailand: Saipunya Samakom.

Sano, Chiaki. 2009. "History of glutamate production." *The American Journal of Clinical Nutrition* 90 (3):728S–732S.

Shahidi, Fereidoon, and Priyatharini Ambigaipalan. 2015. "Phenolics and polyphenolics in foods, beverages and spices: Antioxidant activity and health effects – A review." *Journal of Functional Foods* 18:820–897. doi: https://doi.org/10.1016/j.jff.2015.06.018

Sinsawasdi, Valeeratana K. 1998. "Total Antioxidant Activity in Grape Juice and Spices Products." MS Food Science, Food Science and Human Nutrition, University of Hawaii.

Sloan, A Elizabeth. 2019. "A new wave of Asian Cuisine." *Food Technology*, May 1, 73: 5.

Smith, David V, and Robert F Margolskee. 2001. "Making sense of taste." *Scientific American* 284 (3):32–39.

Spector, Dina. 2012. "Here's What Astronauts Will Eat When They Head to Mars." Business Insider. https://www.businessinsider.com/nasa-mars-menu-2012-7/lightbox?r=AU&IR=T.

Spence, Charles. 2015. "Just how much of what we taste derives from the sense of smell?" *Flavour* 4 (1):1–10.

Spence, Charles. 2018. "Why is piquant/spicy food so popular?" *International Journal of Gastronomy and Food Science* 12:16–21. doi: https://doi.org/10.1016/j.ijgfs.2018.04.002

Spence, Charles. 2020. "Chapter 10 – Multisensory flavor perception: A cognitive neuroscience perspective." In *Multisensory Perception*, edited by Krishnankutty Sathian, and Vilayanur Subramanian Ramachandran, 221–237. Cambridge: Academic Press.

Sriket, Pornpimol. 2014. "Chemical components and antioxidant activities of Thai local vegetables." *Current Applied Science and Technology* 14 (1):18–23.

Sriraja Panich. 2021. "Sriraja Panich the Test of the Original." Sriraja Panich. https://www.srirajapanich.co.th/heritage.php?lang=_en.

Stubbs, RJ, AM Johnstone, N Mazlan, SE Mbaiwa, and S Ferris. 2001. "Effect of altering the variety of sensorially distinct foods, of the same macronutrient content, on food intake and body weight in men." *European Journal of Clinical Nutrition* 55 (1):19.

Sundaravej, Apichai. 2021. "Thailand's Dipping Legend: Sriracha Sauce – Bangkok 101." bangkok101.com, last modified 2021-05-29. https://www.bangkok101.com/thailands-dipping-legend-sriracha-sauce/.

Tang, Claudia S, Vicki W K Tan, Pey Sze Teo, and Ciarán G Ford. 2020. "Savoury and kokumi enhancement increases perceived calories and expectations of fullness in equicaloric beef broths." *Food Quality and Preference* 83:103897. doi: https://doi.org/10.1016/j.foodqual.2020.103897

Trachootham, Dunyaporn, Shizuko Satoh-Kuriwada, Aroonwan Lam-ubol, Chadamas Promkam, Nattida Chotechuang, Takashi Sasano, and Noriaki Shoji. 2017. "Differences in taste perception and spicy preference: A Thai–Japanese cross-cultural study." *Chemical Senses* 43 (1):65–74.

Tsao, Rong. 2010. "Chemistry and biochemistry of dietary polyphenols." *Nutrients* 2 (12):1231–1246.

Tuttle, Brad. 2015, April 16. "Why You Should Blame Millennials for Spicy Fast Food." *Money*. https://money.com/fast-food-spices-sriracha-ghost-pepper-trends/.

WINA. 2019. "Instant Noodle at a Glance." World Instant Noodle Association. https://instantnoodles.org/en/noodles/report.html.

Xoomsai, Terb. 2012. *Rattanakosin Dishes 1982*. Bangkok, Thailand: Saipunyasamakhom.

Yamaguchi, Shizuko, and Kumiko Ninomiya. 2000. "Umami and food palatability." *The Journal of Nutrition* 130 (4):921S–926S.

Yamaguchi, Shizuko, and Chikahito Takahashi. 1984. "Interactions of monosodium glutamate and sodium chloride on saltiness and palatability of a clear soup." *Journal of Food Science* 49 (1):82–85.

Yeomans, Martin R, Laura Weinberg, and Sarah James. 2005. "Effects of palatability and learned satiety on energy density influences on breakfast intake in humans." *Physiology & Behavior* 86 (4):487–499.

Zhao, Cindy J, Andreas Schieber, and Michael G Gänzle. 2016. "Formation of taste-active amino acids, amino acid derivatives and peptides in food fermentations – A review." *Food Research International* 89:39–47.

4

CHEMICAL AND FUNCTIONAL PROPERTIES OF NOTABLE INGREDIENTS

VALEERATANA K. SINSAWASDI, HOLGER Y. TOSCHKA, AND NITHIYA RATTANAPANONE

Contents

4.1 Introduction

Traditionally, Thai people eat three meals and one snack a day. Rice was served in every meal, including snacks. Breakfast is the lightest, usually porridge or rice with clear soup. Lunch has more fat and protein content but is served with minimal accompaniment, or sometimes

DOI: 10.1201/9781003182924-6

just a one-plate meal, e.g., fried rice. The dinner is the most complicated with various menu items (more detail is described in Chapter 6).

The availability of salt and fish, both freshwater and saltwater
fish, depends on the region and has given Thai people an advantage
of developing seasoning. In addition to saltiness, the sauces provide a taste of umami and the aroma of fish fermentation. The pungent smell is usually unpalatable to foreigners, but it is an essential
ingredient and condiment most Thai people crave when Thai food
is unavailable.

Then, influenced by foreign traders, exported ingredients such as
spices were mixed with local ingredients and became a wide range of
herbs and spices mixtures used practically for an endless possibility of
recipes. In addition, coconut cream is added to herbs and spices paste
to become curry, another Thai food identity.

Due to different climates and geographical locations, ingredients
grown or produced in each area have different sensory attributes and
are suitable for a specific dish. Therefore, there are usually varieties
of ingredients used in one kitchen for other purposes. For example,
jasmine rice from *thung-kula-ronghai* for aromatic cooked rice, *keow-
ngoo* sticky rice from Chiang Rai for desert, large dried chili with
intense red color but not very hot from *Bangchang*. However, the fresh
ingredients, such as varieties of vegetables, are seasonally available, so
their uses are not as specific. For example, in a classic Thai cookbook
first published in 1935, a list of vegetables indicates an astonishingly
wide variety of vegetables. On the list are 78 edible leaves and young
leaf shoots, 39 edible tubers and roots, 40 edible flowers, 22 edible
pods with young seeds or legumes, 82 fruits (botanical ovary of plants
that are eaten as vegetables, e.g., tomatoes), and 16 of Chinese vegetables. In addition, each has a recommendation on whether it should
be eaten raw, cooked, or pickled (Cholamarkpijarn 1980).

4.2 Rice

Rice (*Oryza sativa* L.) has long been the staple food of Thai people. Archeology evidence showed that rice, fish, and salt have been
available in Thailand since the prehistoric era (Weber et al. 2010;
Yankowski, Kerdsap, and Chang 2015). Thai rice is not as soft and
sticky as Japanese short-grain rice and not as dry and fluffy as Indian

long-grain rice. Thai jasmine rice or Hom Mali rice is among the thousands of Thai rice varieties. Its unique property is its aroma, which is described as being similar to pandan leaves. The grains are long, and the tips curl upward and appear translucent and glossy. The Khao Dawk Mali 105 (KDML 105) variety was discovered in 1945 and was named and distributed to farmers in 1959. The Hom Mali fragrant rice has surpassed other varieties in terms of higher yield and better fragrance. When cooked, the Hom Mali rice has a characteristic of long-grain rice, i.e., having soft texture, pleasant aroma, low amylose content, low degree of gelatinization, and consistency of a soft gel (Foreign Office of the Government Public Relations Department Thailand n/d; Vanavichit et al. 2018).

Thai Hom Mali rice emits its best aroma when the crop is freshly harvested, and the cooked rice is eaten while still warm. Preferably, the rice grains should be tender and have the right amount of stickiness, which means that the grains are held together loosely without being too mushy. For many years, Thai Hom Mali rice has won the World's Best Rice Contest, i.e. in 2009, 2010, 2014, 2016, and 2017. It was called Thai jasmine rice in 2009 and 2010, but after Cambodia Jasmine rice won in 2012, Thai jasmine rice was renamed Thai Hom Mali (The Rice Trader 2019).

According to the Thailand Standard for Thai Hom Mali rice (Ministry of Agriculture and Cooperatives, National Bureau of Agricultural Commodity and Food Standards 2003), Thai jasmine rice needs to possess a natural fragrance. In addition, the cooked rice texture has to be tender. The tenderness of cooked rice is primarily determined by the proportion of starch component called amylose. It is mandatory that this rice must have amylose content from 12% to 19% at a 14% moisture level. The average length of the grain should be more than 7.0 mm and not less than 3.0 mm wide.

The Thai Minister of Commerce tried to register a patent for Thai jasmine rice as this type of rice is well known worldwide and originated in Thailand. However, the Intellectual Property Department has ruled that the jasmine rice gene is wild rice that does not have a specific geographical origin. Hence, the term used for its seal of approval is "Thai Hom Mali Rice – Originated in Thailand." Further details on this topic can be found on the Department of Foreign Trade Regulations website at http://www.thai-hommalirice.com.

4.2.1 Starch in Rice

Rice grains can be milled and polished to remove the bran and germ, leaving only the white starchy endosperm part in the middle. The white endosperm part is considered "refined grain," which means it does not provide the benefits of dietary fiber. Thus, the glycemic index is high as the refined grain can be digested and absorbed very quickly. Unpolished or brown rice is considered "whole grain" and is a good choice for those who want to increase their whole-grain intake.

The endosperm, the largest part of the grain, is starchy and contains starch granules. Each granule contains several millions of polysaccharide molecules. Starch polysaccharide consists of a long chain-link of several hundreds of glucose units, so polished rice is easily digested to glucose. Though all the polysaccharides in rice starch contain just glucose molecules, there are two types of arrangement. The linear one is called amylose, and the other branched one is called amylopectin. The proportional difference in amylose and amylopectin produces a wide range of textures among the different varieties of rice when cooked (Wong 2018a).

4.2.2 Gelatinization of Starch in Rice Grain

In natural forms, such as in rice grains, the amylose and amylopectin are packed so tightly that the digestive enzymes in humans cannot reach and break down the bond between the glucose molecules. In other words, uncooked rice is inedible and indigestible. Water is needed to hydrate the amylose and amylopectin, but energy, in the form of heat, is also required for the water to be taken into the granule (imbibition). As rice grains take up water with heat, the starch granules inside will swell, and some small amylose molecules can leach out of the granules. This crucial step in cooking rice is called gelatinization (Eggleston, Finley, and deMan 2018). Once fully gelatinized, cooked rice is soft and palatable. However, once cooled, the starch polysaccharides will start to lose the ability to absorb water and begin to align themselves into densely packed grains again. When the cooked rice becomes hard, this process is called retrogradation or staling. Cooked rice with a mild degree of retrogradation is suitable for making fried rice because the grains will stay separate and firm, which is the desired quality expected from a fried rice dish.

The type of starch molecule and the degree of gelatinization also have important nutritional implications. Starch that is fully gelatinized takes less time to digest and absorb. Thus, blood glucose will rise rapidly after consumption. A measurement of how fast blood glucose increases after the consumption of carbohydrates is called the glycemic index or GI. The glycemic index has a scale of 0–100, with a glycemic index of over 70 considered to be high, while a value of less than 55 is considered low. Freshly cooked rice has a high glycemic index because gelatinized amylose and amylopectin bind with a lot of water molecules. During cooling and storage, retrogradation occurs. With the closer alignment of starch molecules, the digestion of starch to release glucose into the bloodstream is delayed, so stale rice tends to have a lower GI. Since starch with a higher amylose content tends to undergo a greater degree of retrogradation during storage, high-amylose rice has a lower GI (37–64).

Another factor affecting the GI of rice is the amount of fiber in the rice grain. Unpolished or brown rice retains the bran layers of the rice grain and, in the stomach, the fiber molecules interfere with the enzymatic digesting of starch and thus delay the release of blood glucose. Brown and unpolished rice, therefore, have a lower GI than white rice, with GI ranging from 50 to 87 (Foster-Powell, Holt, and Brand-Miller 2002; Kwanbunjan 2020). Therefore, those who want to control blood sugar for diabetes prevention purposes should opt for lower GI type of rice, especially brown rice.

4.2.3 Characteristics of Cooked Rice

Hom Mali rice has an amylose content of between 12% and 19%. This amount is relatively high compared with short-grain rice, which has less than 5% of amylose. In general, the higher the proportion of amylose, the higher is the firmness of the cooked rice grain. The acceptability and palatability of cooked rice greatly depend on the degree of hardness, stickiness, and moistness to the touch. In detail, the firmness of cooked rice has a positive relationship with the length of the linear part of both amylose and amylopectin (Li et al. 2016; Ramesh, Bhattacharya, and Mitchell 2000).

With the high proportion of amylose in Thai rice, the grains are firm. In contrast to long-grain rice, another popular type consumed

primarily in the north and northeastern parts of Thailand is glutinous or sticky rice (*khao-neow*, ข้าวเหนียว). Sticky rice is considered a
short grain and contains mainly amylopectin (less than 5% of amylose). With lower amylose, the stickiness of the cooked rice is high.
With the grains sticking together, fluffing the rice is not as necessary.
Stickiness can also be caused by amylose and amylopectin leaching
out from the cell.

Cooking rice method: In those days, before the electric rice
cooker became available, there were two primary traditional cooking methods for rice. Both methods required water and heat to boil
the rice, but one method uses a lot of water and involves pouring the
boiling rice water away toward the end. The other one requires only
just enough water (1–2 times of the rice volume), with no drainage
of the excess water. After the rice has been boiled until almost fully
cooked (still leaving a rigid core, somewhat harder than the *al dente*
pasta cooking), the heat is turned off, and the rice is left in a tightly
closed container for about 10 minutes to let it become fully cooked.

The boiling step is to gelatinize the starch granules in the rice
grains. Starch absorbs water and swells at a temperature of around
65°C. Stirring is required only occasionally to avoid rupturing the
starch granules, reducing the rice grains' firmness. The last step,
which requires only a little or no heat source, is to allow water uptake,
homogenously, in every starch granule. The technique is known as
dong-khao (ดงข้าว). This practice also helps to avoid burning on the
bottom and the sides of the cookware. When there is water around
the grains, the energy goes to heat up the water and the maximum
temperature throughout the process is 100°C that is the boiling point
of water. If the heat is still high, once the water has all been absorbed
into the rice grains, the energy will heat the grains to over 100°C,
risking overcooking or burning.

Though it largely depends on personal preference, rice is usually
fluffed up after cooking. This step is known as *sui-khao* (ซุยข้าว), or *sai-
khao* (สายข้าว) in case of glutenous rice. The fluffing can be done when
the rice is freshly cooked, after resting it for about 5–10 minutes.
Then, to allow the steam to dissipate, use a large spatula or rice
paddle to gently flip the rice from the bottom to the top without
pressing and continue to gently mix until the rice is aerated and
contains no lump.

Almost every household uses an electric rice cooker to cook rice, but the same principles still apply. The process starts with adding rice and water to the inner cooking bowl of the rice cooker. The amount of water depends on the rice type and storage time and is usually specified on the rice packet. The heating continues until all the water has been absorbed into the rice grains. At this point, the thermocouple, which is a device that measures the temperature, detects that the temperature is now rising above the boiling point of water (100°C) and will stop heating. The cooked rice should be left in the cooker for 5–10 minutes, then fluffed up if desired.

The amount of water absorbed into rice grains during cooking may differ with the rice's variety and storage period. Rice stored at higher temperatures tends to have a lower water absorption capacity over time, resulting in an increased hardness of the cooked rice. During storage, starch rearrangement results in stronger bonding. Also, there is an increasing crosslinking of hemicellulose into stronger cell walls. These changes may contribute to cooked old rice's decreased water absorption and increased hardness. Newly harvested rice tends to leach starch molecules out of the granules, so the cooked rice is stickier and lumpier. Aged rice tends to be less sticky and fluffier (rice grains swell but still detach from each other, leaving some space in between). If rice is stored at ambient temperature, it tends to lose flavor, texture, and appearance and is at risk of rejection due to the increased odor of free fatty acids from lipid oxidation. Aroma compounds in rice decrease as the rice ages. Better packaging, such as in laminated polypropylene, aluminum foil, or polyethylene, is efficient in preventing the loss of aroma compounds and delays lipid oxidation (Parnsahkorn and Langkapin 2013; Ramesh, Bhattacharya, and Mitchell 2000; Tananuwong and Lertsiri 2010; Tananuwong and Malila 2011).

During cooking, rice grains absorb water at about 2 g of water per gram of rice. Therefore, the volume of water added until just covering the rice in any container used is just about the volume absorbed into the rice grains. However, because of the long boiling period, water will be lost in the form of steam. The loss is roughly proportional to the surface area of the container, so extra water is needed to compensate for this loss. The most simple and classic method Thai people use to judge the amount of water is by measuring the water level with a finger. Simply touch the rice with a finger (middle or index), and the water level should reach the finger's first knuckle. This method works very well for household cooking, probably because the differences in the length of this first segment of a finger are about 2.5 cm and do not vary much between people (Peters, Mackenzie, and Bryden 2002).

For steaming rice, as in the case of sticky rice cooking, it is not necessary to compensate for water loss because the rice is not immersed in water, and steam is continually generated from the saucepan under the rice. However, to ensure that each grain absorbs enough water and accelerates the starch's gelatinization, the rice must be soaked in water until it is fully absorbed before putting it in a steamer. In addition, if the rice is newly harvested and the cooked rice is too sticky, rinsing the rice a few times before cooking can help wash away some extra starch outside the grains.

4.3 Fish Sauce (*Nam-Pla*)

Among the ten countries in Southeast Asia (members of the Association of Southeast Asian Nations or ASEAN), the use of seasoning and condiments derived from the fermentation of fish and shellfish has been identified as a common feature of ASEAN food culture. Depending on geographical location, either freshwater or saltwater fish can be used. For example, there are similar freshwater fish fermentations such as Thai *pla-ra* (ปลาร้า), *Prahok* in Cambodia, *Ngapi* in Myanmar, and *Mum* in Vietnam. Also, there are similar saltwater fermentations such as Thai *nam-pla* (น้ำปลา), *teoktrey* in Cambodia, *nam-pa* in Lao, *kecap* in Indonesia and *patis* in the Philipinnes (Panyayong 2016).

Thai fish sauce or *nam-pla* (น้ำปลา) is the main ingredient that gives a salty taste and unique aroma to most savory dishes. Fish sauce is a common condiment on the dining table, the same concept as the salt shaker in Western culture. The flavor of the fish sauce itself can be described using sensory attributes such as sweet, caramelized, fermented, fishy, and musty to describe the odor; sweet, salty, bitter, and umami to describe the taste; and caramelized and fishy to describe the overall flavor. Interestingly, sniffing (through the orthonasal route or by breathing in) fish sauce results in a weak sweet aroma along with a less desirable fermented and musty effect. However, once the fish sauce is in the mouth, the distinct flavors perceived (through the retronasal route or by swallowing) are caramelized and fishy. The overwhelming taste is salty and umami, with a distinctly salty after-taste (lessening the degree of the sweet and umami aftertaste). The key characteristics of Thai fish sauces are a fishy aroma, a fishy flavor, with a sweet aftertaste (Ritthiruangdej and Suwonsichon 2007).

As the name implies, the sauce is made from the fermentation of fish and salt. Anchovies or small sardines (*Stolephorus* or *Sardinella* species) are used in the production. Generally, there will be 2–3 parts of anchovy for each part of salt, depending on the recipe, usually in a large concrete tank. Next, the mixture of salt and fish is just left for natural microorganisms to start the fermentation process, which takes about a year to a year and a half. Finally, the fish is discarded, and the liquid obtained from the fermentation will be filtered, bottled, and ready to use.

During fermentation, proteins and lipids (fats) from the fish are degraded or hydrolyzed by enzymes and bacteria. The breakdown of fats results in various short-chain free fatty acids that have a small molecular size, so human olfactory receptors can detect them. These small chemicals or molecules that produce the odor are called volatiles, volatile substances, or aroma components.

There are aroma components in the fish sauce that are derived from protein. Protein in the anchovies is hydrolyzed or broken down by enzymes. This group of enzymes is called protease because of its specific function in breaking down protein. The protease is found naturally in fish muscle and the digestive tract. Bacteria capable of thriving in a high salt environment, such as in fish sauce, are called halophilic bacteria. These bacteria also produce protease enzymes that split the bonds in protein resulting in smaller fractions of the protein chain of peptides and free amino acids. The breaking down or hydrolysis of protein and lipids go on throughout the months of fermentation until the distinctive flavor and odor of fish sauce have developed. Following this principle, instead of relying on the natural protease enzymes from the fish and natural microorganisms, it is possible to select strains of bacteria that contain the desired protease that can be added to the salted fish to obtain the desired protein hydrolysates and reduce the fermentation time from 12 months to 4 months (Saisithi 1994; Yongsawatdigul, Rodtong, and Raksakulthai 2007).

Scientists can identify the molecular structure of these aroma components by first separating them using gas chromatography. Then, the molecular structure of each isolated compound can be identified further by mass spectrometry, and almost simultaneously, a human can characterize their smell. This technique is called gas chromatography-olfactometry (GC-O), and this technique primarily identifies the smells of other foods and ingredients mentioned throughout this book.

The GC-O technique is used with many examples of fish sauce in many studies. The odor of fish sauce can be described in several terms, from positive to negative food-related descriptions. These terms are ammoniacal, cheesy, meaty, burnt, fishy, sweaty, fecal, rancid, nutty, malty, potatoey, brothy/meaty, fishy, briny, sulfury, seashore-like, crab meat, cabbage, sweet, ethereal, flowery, sweet, boiled rice-like, coffee-like, metal, paint, fruity, garlic-chive-like, fatty, oily, rusty, plastic, strong, pungent, sour, vomit-like, medicinal, and chemical (Lapsongphon, Yongsawatdigul, and Cadwallader 2015; Saisithi 1994; Wichaphon et al. 2012). With fish sauce being produced throughout Southeast Asia, its aroma tends to be pretty varied.

In general, in households, the preference for fish sauce characteristics seems to be varied, and the brand of fish sauce plays a significant role in the criteria for buying. Its nitrogen content officially grades the quality of the fish sauce. The first grade contains 20 g of nitrogen per liter, while the second one contains 15 g of nitrogen. Though not officially recognized, premium grade typically has more than 20 g of nitrogen per liter, which indicates a longer fermentation time and the higher concentration of aroma compounds.

It is not only the saltiness and richness of aroma provided by fish sauce but also the enhancement of the meaty and savory taste of umami in a dish. The active compound responsible for the umami taste is glutamate that is the product of protein hydrolysis. There are many amino acids found in fish sauce, and the most common ones are glutamic acid, lysine, aspartic acid, histidine, and proline. More details can be found in Chapter 3 of this book.

4.4 Other Salty Seasonings

Though fish sauce has been recognized as the main seasoning that achieves saltiness and flavor, there are many other seasonings in other regions used for the same purpose but in different dishes. According to the national survey, the most popular seasoning for Thai people is fish sauce, salt, fermented shrimp paste (*ga-pi*, กะปิ), soy sauce, and the relatively new varieties of commercial seasoning sauce (*sauce-proong-ros*, ซอสปรุงรส) (National Bureau of Agricultural Commodity and Food Standards, Ministry of Agriculture and Cooperative 2019). The seasonings that involve fish fermentation are *bu-du* (บูดู) and *pla-ra* (ปลาร้า). See Figures 4.1 a-g for pictures of these seasonings. *Bu-du* is fermented

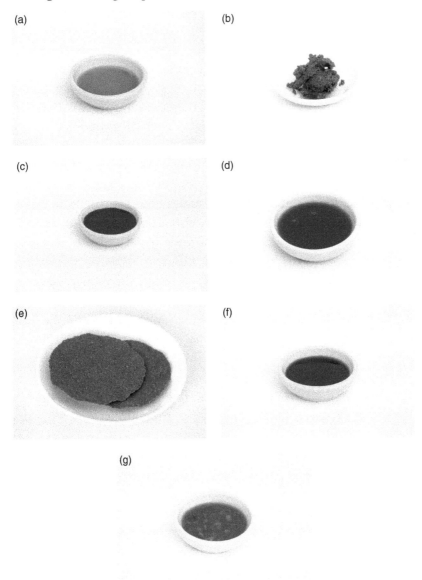

(a)

(b)

(c)

(d)

(e)

(f)

(g)

Figure 4.1   (a) Nam-pla. (b) Ga-pi. (c) Pla-ra. (d) Bu-du. (e) Tua-nao. (f) Soy sauce. (g) Tao-jeow.

from saltwater fish just like fish sauce and is used in the southern part of Thailand, but the liquid is not as transparent. *Pla-ra* is popular in the north and northeastern parts, far away from the seashore, so the fish used in this condiment is various freshwater fish.

Besides *pla-ra*, there are also condiments made from soybean and another from freshwater crab in the north. The fermentation of soybean that is quite similar to the Japanese natto is called *tua-nao* (ถั่วเน่า). *Tua-nao* can be dried and grounded into a powdered seasoning for northern-style soups and curries. *Nam-poo* (น้ำปู) is an extract of crab mixed with herbs such as lemongrass and galangal. Lastly, *ga-pi* (กะปิ) is fermented shrimp paste, a product that is a common condiment throughout the country (Chotechuang 2012). The most popular sauce from this shrimp paste is the main component of chili paste or *nam-prig-ga-pi* (น้ำพริกกะปิ). Though all these seasonings share similar characteristics of high salt content, their aroma profile and glutamic acid content for umami taste enhancement vary. In addition, each product provides a unique aroma to a specific food it is traditionally paired with. Hence, they are not substitutable without noticing. With a wide variety of raw materials, protease enzymes are provided by different types of culture and different fermentation processes. As a result, the fermented products are varied in taste and smell. Bacterial cultures and enzymes for fermentation are endogenous to the raw materials and the local environment, especially halophilic bacteria that already exist in the internal fish organs. Hydrolysis of protein and nucleic acids yields glutamate and nucleotides, the basic umami taste substances and the source of deliciousness in dishes. The microbial cultures used in these seasoning are, for example, *pla-ra* from fermented freshwater fish fermented with *Salinivibrio siamensis* and *Virgibacillus siamensis*. *Ga-pi* is the fermentation of planktonic shrimp (Genus Acetes and Mesopodopsis) with bacteria, such as one that was isolated and proposed to be named *Virgibacillus kapii*. *Nam-bu-du* or fish sauce is obtained from saltwater fish and *Bacillus subtilis*, and *tua-nao* from fermented soybean with *Bacillus subtilis* (Ajinomoto Academic Department 2011; Choorit and Prasertsan 1992; Chukeatirote 2015; Dajanta et al. 2011; Daroonpunt et al. 2016; Hajeb and Jinap 2015; Suriyanont and Chancharoonpong 2018; Tanasupawat et al. 2010).

All these seasonings and applications in several dishes in all regions throughout the kingdom result from local pearls of wisdom that have helped develop the character of authentic Thai cuisine. Though they are similar in providing saltiness and umami tastes, they are rarely interchangeable because the differences in volatile compounds result in the vastly different flavor profiles needed explicitly for each dish. For example, **tom-yum-koong** requires *nam-pla* (fish sauce), **khao-yum** requires *nam-bu-du* (southern-style fish sauce), **som-tum-pla-ra** requires *pla-ra* (northeastern-style fish sauce), **khao-klook-ga-pi** requires *ga-pi* (fermented shrimp paste), and **nam-prig-ong** requires *tua-nao* (dried fermented soybean paste). General recipes for these dishes are provided in Chapter 8.

There are many types of seasoning from fermented soybean. Chinese-style soy sauce is fermented solely from soybeans. The Japanese style, on the other hand, is a combination of soybean and wheat. Regardless of the bean sources, these seasonings serve as flavor enhancers because they contain umami taste substances, mainly glutamate (Lioe, Selamat, and Yasuda 2010). Soy sauce is used in recipes that originated in China, such as noodle soup, five-spice egg soup (*kai-pa-lo*, ไข่พะโล้), stir-fried vegetables, stir-fried noodles (*pad-see-ew*, ผัดซีอิ๊ว) but not in traditional authentic Thai foods.

Not all salty seasoning available in Thailand is a result of fermentation. There is also a relatively new seasoning sauce, such as the Maggi sauce famously industrialized by Julius Maggi. This type of sauce uses acid to hydrolyze bean or wheat (instead of microbial fermentation) to provide the meaty taste in soup (Giacometti 1979). This type of sauce is now known as hydrolyzed vegetable protein (HVP). The non-fermented soybean-based sauce has been developed into a variety of seasoning sauces.

The traditional salty, savory seasoning fermented from a single main ingredient such as fish sauce and soy sauce has gradually been replaced by varieties of the generically called seasoning sauce (*sauce-proong-ros*, ซอสปรุงรส). Most of the seasoning sauce is in liquid form packaged in a bottle. The texture varied from very watery to very viscous one like the oyster sauce. With the power of marketing, these sauces are commonly referred to by commercial brands. The main ingredients of these seasonings are still mostly soy sauce or HVP but with added MSG, sugar, artificial color, artificial flavors, and preservatives.

Oyster sauce was a relatively new addition to the Thai kitchen. It was initially imported to Thailand from China before the first oyster sauce factory in Thailand was founded in the 1970s. The sauce not only gives flavor and saltiness but also provides brown color and viscosity. The most common dish with oyster sauce is stir-fried vegetables.

4.5 Coconut Milk

Coconut (*Cocos nucifera* Linn.) is abundant in Thailand and is highly beneficial to the Thai people. Local Thais prefer to drink coconut water from the young coconut fruit with scraped-off thin, translucent coconut meat. When the fruit is more mature, the meat or the endosperm will thicken and contain a high amount of fat, the source of coconut milk. In addition, coconut flowers produce syrup in sufficient quantities, which is processed into coconut palm sugar, simply by heating to evaporate the excess water.

Coconut milk is used in both savory dishes and desserts. Thai people used coconut milk to replace ghee and butter when Indian curry was introduced to Siam because dairy products were not available in the old days. In the late twentieth century, heat processing and packaging technology yielded shelf-stable products such as pasteurized coconut milk in a plastic bag, sterilized coconut milk in a metal can, and ultra-high temperature (UHT) with aseptic packaging. These products have almost totally replaced the traditional method of fresh coconut milk preparation.

The outer layer of coconut is a husk, which becomes fibrous and turns dark brown once mature. The young coconut has a soft spongy husk with a very light color, and the outside is green. The shell is next to the husk, turning from light to dark as the fruit matures. Inside the coconut is a cavity composed of two types of endosperm; the solid one is the coconut meat or kernel that gains a higher fat content as it matures, and the liquid one is the coconut water.

Young coconut fruit can be served as a drink with sweet coconut water and light, sweet, thin, and soft coconut meat. It takes about 12 months for a coconut to become fully mature. To estimate the maturity of the coconut fruit, one can shake the fruit and listen to the sound of liquid inside. Young coconut will be filled with the juice so the shaking does not make a noise. However, the mature one contains less fluid, so shaking the fruit will produce a noticeably loud noise.

Mature coconut has thick meat and contains 20-44% fat. The stiff and rigid coconut meat is scraped from the shell using a handheld grater or a grater attached to a stool to prepare fresh coconut milk manually. Artisans in the past carved the shape of the small seat with the attached steel grater into a rabbit, so this classic tool is called a "coconut grater rabbit" or *gra-tai-kood-ma-prow* (กระต่ายขูดมะพร้าว)

In home cooking, the first squeeze of coconut milk is called *hua-ga-ti* (หัวกะทิ), directly translated as the head of coconut milk. It is compatible with cream as it is the most concentrated part with the highest fat content and can thus be called coconut cream. A cook simply mixes the grated mature coconut meat with warm water, up to a ratio of 1:1, and repeats the motion of "squeeze and release" in a bowl until the coconut milk is visible (Fig. 4.2a and 4.2b). The coconut milk is filtered with several cheesecloth layers. This coconut cream (*hua-ga-ti*) is kept separately since many recipes specifically require this first squeeze or heavy cream.

The remaining coconut meat can produce more coconut milk by simply repeating the process of the first squeeze but using more warm

Figure 4.2   (a) Grated mature coconut meat. (b) Manual extraction of coconut milk. (c) Freshly squeezed coconut milk, hua-ga-ti (left) and hang-ga-ti (right). (d) Hua-ga-ti (left) and hang-ga-ti (right) 50 minutes after extraction.

water in a ratio of about 1:2. Depending on how much fat content in meat, the process can be repeated for up to three times. The coconut milk obtained from the second and third squeeze is much thinner with a much lower fat content. It is called *hang-ga-ti* (หางกะทิ) implying that this portion is a lighter cream and coconut milk, roughly equivalent to low-fat milk. Simply dilute down with water to the desired consistency when using processed coconut milk (canned or UHT).

There have been some investigations into the best conditions for the efficient extraction of coconut milk from coconut meat. The ratio of coconut meat and water does not matter, but the water temperature is essential. Extraction at 55°C yields the highest fat content compared to 30 and 80°C. The preparation method of squeezing without water first, then following with squeezing with water or squeezing twice with an equal amount of water or just once with all water, does not yield a different amount of fat (Patil and Benjakul 2018).

If coconut milk is to be used in curry dishes, the dark color of coconut milk is acceptable since it will eventually become dark in the dish anyway. On the other hand, desserts are almost all light in color and require totally white coconut milk. So, for desserts, the process has to start from the grated meat with no trace of the brown coconut shell. Since the yield is different, this type of coconut milk is more expensive.

Like cow milk, coconut milk appears to be white and opaque because of the emulsion system of fat globules dispersed in water. Fat globules, which are dispersed throughout the milk, block the passage of light the same way as in cow's milk, giving a milky appearance. Typically, oil and water cannot mix (they are immiscible). Still, coconut milk contains protein globulins and albumins, along with phospholipids surrounding the oil droplets that help to prevent these oil droplets from merging and rising to the top. Thus, in chemistry terms, coconut milk is an oil-in-water emulsion with coconut proteins (globulins and albumins) and phospholipids as emulsifiers (Chiewchan, Phungamngoen, and Siriwattanayothin 2006; Patil and Benjakul 2017; Tangsuphoom and Coupland 2005).

Due to the size of the fat globules in natural coconut milk and relatively low emulsifiers, coconut milk tends to separate into a creamy layer on top of the translucent water layer at the bottom within an hour of storage (Fig. 4.2c and 4.2d). In the food processing industry, coconut milk undergoes homogenization to help reduce the size of fat globules. Though emulsion particle size becomes smaller after homogenization, the creaming or separation of a fat layer from the water layer is apparent due to the relatively low emulsifying protein. Therefore, some manufacturing plants may add chemical food additives, especially emulsifiers and stabilizers to increase physical stability and prolong the product shelf life (Ariyaprakai, Limpachoti, and Pradipasena 2013; Tangsuphoom and Coupland 2009).

There are many types of commercial shelf-stable coconut milk available in many formats, such as pasteurized coconut in a pouch, UHT processed, canned, and powder. Changes in the aromatic compounds, such as fatty alcohol and small fatty acid molecules, occur at over 75°C (oxidation to aldehyde and ketone substances). Then, if the processing temperature is above 85°C, sugar degradation and the Strecker degradation of amino acids may occur. As a result, the characteristic flavor associated with freshly squeezed coconut milk is lost. Hence, while fresh coconut milk is associated with fresh odor, high heat processed products like UHT, canned, and spayed-dried powder have more aroma attributes of coconut oil, cooked, nutty, and sweetness compared to fresh and pasteurized coconut milk (Wattanapahu et al. 2012).

For home use, whether freshly prepared coconut milk or processed, the separated phases can be mixed into an emulsion, at least temporarily, by simple shaking. However, if the coconut milk is to be used in curry, it is recommended that a can of coconut milk be opened without prior shaking. The existing separation will allow for the coconut cream layer to be scooped out of the can easily. In addition, the higher fat concentration in the coconut cream reduces the time needed to heat until oil is released from the cream. Hand-squeezed coconut milk takes less heating time because it has not undergone a homogenization process, and there is no added emulsifier.

When coconut milk is used to make curry, coconut milk emulsion will have to be destabilized by heat. The collapse or breaking down of emulsion yields visible oil. Heat denatures and coagulates coconut milk protein when the temperature reaches at least 80°C (Kwon, Park, and Rhee 1996). With the coagulation of protein surrounding the oil droplets, the molecules lose the ability to act as an emulsifier. The oil droplets then combine (or *coalesce* in chemistry terms) to form larger oil droplets. Eventually, the oil phase will be separated from the water phase. The optimum temperature to destabilize the emulsion and release the oil is 90°C (Raghavendra and Raghavarao 2010).

Adding curry paste is recommended at this stage when the oil becomes highly visible, this stage is called *ga-ti-tag-mun* (กะทิแตกมัน). The heat releases oil from the emulsion and induces a further break-down or partial hydrolysis of fats to create short-chain free fatty acids and other volatile compounds. Thus, accompanying the oil appearance, the odor of coconut milk is much more durable.

Once the heated oil reaches the curry paste, the hot oil further extracts oil-soluble compounds out of the paste, creating the potent smell of sweet and creamy coconut together with the intense and spicy aroma of herbs and spices in curry paste. People around the kitchen usually sneeze at this step due to the burst of aromas. Most volatile compounds, including those from herbs and spices, are hydrophobic. These compounds tend to solubilize in the oil phase. Hence, lipids or oil released from coconut milk acts as a solvent to help extract these hydrophobic flavor components from the herbs and spices in curry paste. With a higher concentration of hydrophobic aroma compounds, the dominant flavor is in the oil phase. The coconut cream has to be heated long enough for the emulsion to break down (*ka-ti-tag-mun*) and the oil layer, called *kee-lo* (ขี้โล้), becomes substantial. If this is achieved, the characteristic flavors of coconut milk via complicated lipolysis can be fully developed.

The non-fat (no coconut milk) version of curry such as jungle curry (*gaeng-pa*, แกงป่า) is less flavorful than the coconut-milk-based type of curry. The difference is due to flavor compounds that are more fat-soluble than water-soluble. Removing or reducing the fat content from food drastically alters flavor profile (Bayarri, Taylor, and Hort 2006). Also, since capsaicin, the compound that gives the hot sensation in chili, is fat-soluble, the flavor and mouthfeel of the no-coconut milk curry are not as smooth and intense as those with coconut milk. The fatty acid profile and characterization of both the fat and emulsion structure affect how flavors are released from the food matrix and how each emulsion contributes to the rate of delivering flavor compounds (Mao et al. 2017; Relkin, Fabre, and Guichard 2004). Therefore, an ingredient with high-fat content, such as coconut milk, contributes to the texture and oral sensation of the food. Also, the development of flavors from the combination of emulsion and flavor compounds such as coconut milk and Thai curry paste is a unique temporal flavor profile (dynamic changes of flavor being released at each step of cooking to chewing in the mouth).

When using coconut for desserts, for example, Banana in sweetened coconut milk or *gluay-buad-chee* (กล้วยบวชชี) to the desired consistency that is thick and creamy. In this case, the release of oil or emulsion breakdown is undesirable. Thus, time and temperature need to be controlled carefully during the cooking process to obtain the desired character of the finished product.

Apart from the appearance of shelf-stable processed coconut milk, the product's aroma is vital for the quality of the final food product. After two months of storage, there is an increase of a molecule called *lactone* that increases coconut-like, creamy, and sweet odors in the canned product. After five months of storage, the perceived aroma remains unchanged, volatile compounds detected are those that give a nutty and roasted cooking odor. The odors that fade after five months are pungent, coconut-like, caramel-like, but, most noticeably, an increase in the toffee-like smell (Tinchan et al. 2015).

Coconut milk is prone to microbial spoilage as it is rich in nutrients that support the growth of microorganisms. The lengthy preparation process, which requires several tools and utensils and the food handler's hands, also contributes to the introduction of microorganisms into the coconut milk. Common microorganisms found in coconut milk are bacteria, such as coliform bacteria and *Bacillus,* and yeast, such as *Saccharomyces* spp. If the coconut milk is left at a warm temperature, i.e., 35°C, the flavor of coconut milk will be greatly altered in just 6 hours. Not only microbial spoilage but also the activity of various enzymes and susceptibility to oxidation also rapidly change the sensory properties of coconut milk (Seow and Gwee 1997).

4.6 Chili Peppers

Chili pepper was originated in South and Central America and brought to Siam by Western traders. The red color of chili is composed of pigments called capsanthin and capsorubin that are referred to as oleoresin. Oleoresin is used mainly for coloring purposes.

(a)   (b)

Figure 4.3 (a) Variety of fresh chili. (b) Variety of dried chili.

The heat and pungency of chili are due to capsaicinoids that have no color nor odor. The main composition of capsaicinoids is capsaicin and dihydrocapsaicin (Aza-González, Núñez-Palenius, and Ochoa-Alejo 2011; Kumar et al. 2011).

In Thailand, two main types of chili are used. One is bird's eye chili or *prig-kee-noo* (Fig 4.3a on the left), and the other type has longer fruit and is called long chili or spur chili (*prig-chee-fah* Fig 4.3a in the middle, พริกชี้ฟ้า). Sun-dried chili (Fig 4.3b) is also used in various dishes, especially in curry paste. Studies of capsaicinoids to assess the degree of pungency measured in terms of Scoville Heat Units (SHU) have found that most Thai chilies are classified in the highly pungent category with a pungency level much higher than 25,000 (SHU – ppm of capsaicinoids × 15). The highest capsaicin content of each chili cultivar is found in the first harvest, meaning the subsequent harvest is less pungent (Kraikruan et al. 2008).

With dry heat such as baking, chilies gradually lose the aroma of raw chili, but certain compounds such as esters are found as baking continues. In terms of pungency, the longer the chili is heated, the less pungent it will become (Srisajjalertwaja et al. 2012). However, when combined with other ingredients such as garlic, shallots, dried shrimp, fish sauce, shrimp paste, and coconut sugar and then fried, aroma compounds become complicated. For example, there is the development of odor described as dark chocolate, butter, floral, and garlic, while odors such as fishy and meaty become weaker with more extended heat (Rotsatchakul, Chaiseri, and Cadwallader 2007).

Although chili was not first cultivated in Thailand, the Thai people have mastered the technique of incorporating chili into many recipes. As a result, the spiciness of chili has become a symbol of Thai food. Further details on chili and its role in Thai cuisine are in Chapters 3 and 6.

4.7 Palm Sugar

The word sugar in Thai is *nam-tarn* (น้ำตาล), which literally means liquid from a palm tree. The English language for "palm" may refer to either the coconut palm or palmyra palm trees, both of which are in the palm tree family (Arecaceae) and capable of producing sap with high sugar content. However, both palms are from different genus. Palmyra palm (*Borassus flabellifer* Linn.) is much taller than coconut palm (*C. nucifera*), making it harder to collect the sap. It is believed that palmyra palm sap was the first source of sugar used in both Thai savory dishes and desserts and this sugar product was called *nam-tarn-ta-node* (น้ำตาลโตนด). Later, when coconut trees became commercially cultivated, their shorter tree with lower hanging fruits made it much easier to harvest the sap. With this significant advantage, coconut syrup has become the predominant source of sugar.

Evidence about coconut cultivation is in a written record dating back to the Sukhothai era (around the thirteenth century). The sugar from the coconut tree is called *nam-tarn-ma-prao* (น้ำตาลมะพร้าว) and is much more widely available commercially to this day (Kongpan 2018). Generally, the term palm sugar was used to refer to either palmyra palm sugar or coconut sugar and was usually defined by the character of the finished product (Figures 4.4 a-b). For example, *nam-tarn-mor* (น้ำตาลหม้อ) and *nam-tarn-peeb* (น้ำตาลปี๊บ) refer to semi-solid palm or coconut sugar stored in a saucepan (*mor*) and on a metal pail (*peeb*), respectively. *Nam-tarn-pueg* (น้ำตาลปึก) describes sugar that is a thick solid mass, approximately the size of a small coffee cup.

The palmyra palm tree can grow up to 30 m tall, and the fruit called toddy palm (*look-tarn*) is usually cooked with sugar to make desserts.

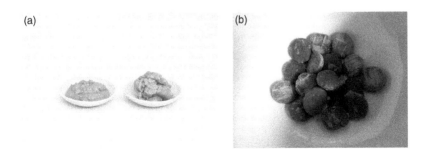

(a)   (b)

Figure 4.4   (a) Palm sugar, concentrated. (b) Palm sugar, dried.

Toddy palms in syrup are commercially available in metal cans. The part of the palmyra palm tree that produces sap is the inflorescences (clusters of flowers). The sap or nectar that flows out of the inflorescences is traditionally collected by bamboo tubes. Usually, there are two rounds of collections a day as the process takes several hours of drop-by-drop collection. To prevent fermentation (spoilage) during these hours, ancient wisdom advises are to add locally grown wood pieces of *Kiam* (เคี่ยม, *Cotylelobium lanceolatum* craig.) or *pa-yorm* bark (เปลือกพะยอม, *Shorea rofburthii* G Don) to the collecting tube. The wood pieces will be filtered out before further processing, i.e., evaporating off the water until the desired concentration, color, and consistency. Without preservation, the sap can undergo several stages of fermentation, resulting in lactic acid, alcohol, and acetic acid dependent on the microbial load, time, and temperature (Saputro, Van de Walle, and Dewettinck 2019).

The sugar content of the sap ranges between 10-17%, and the majority of the sugar is sucrose. The ratio of fructose: glucose: sucrose ranges from 1:1:9 to 1:1:21. However, the glycemic index of this sweetener is low compared to the regular kitchen sugar (purely refined sucrose). The lower glycemic index is possibly because both types of palm sugars (palmyra palm and coconut palm) contain other constituents such as amino acid, vitamins, and minerals. Also, there may be some non-starch polysaccharides like inulin (Srikaeo and Thongta 2015; Trinidad et al. 2010).

The sap is very aromatic and can be served as a beverage on its own. Distinct characteristics of the sap are not only its sweetness but also its fragrance. Volatile compounds in the palmyra palm sap, such as varieties of ketones and esters, produce the attributes of fruitiness, citrus, cheesiness, butteriness, sweetness, and the herbal notes to the product. When heated, the number of volatile compounds decreases, especially 3-hydroxy-2-butanone (sweet aroma) and 1,3-butanediol (buttery aroma), but there is a rise of 2,3-dihydrobenzofuran (giving the smell of caramel) (Naknean, Meenune, and Roudaut 2010; Saputro, Van de Walle, and Dewettinck 2019).

Protein content in the palmyra palm sap is up to around 0.4 mg/g. The amino acids detected are, for example, glutamic acid, threonine, aspartic acid, and serine. With the reduction of both sugars (fructose and glucose) and the amino acid content, the syrup prepared from the heating of sap undergoes the Maillard reaction upon heating to get syrup or sugar cake. At temperatures higher than 120°C, a caramelization reaction takes place. Both Maillard and caramelization reactions are responsible for the darkening of the syrup during palm sugar production. The amino acids in palm sugar and the abundance of sucrose enable further chemical reactions such as Strecker degradation, forming several more volatile compounds. The traditional method of heating is in an open pan with constant stirring. The flavors associated with palm sugar, such as roast, nutty, sweet, and caramel, are fully developed when the processing temperature reaches 110°C.

Besides an increase in brown color intensity, there is also the development of hydroxymethylfurfural (HMF) as a specific product from the Maillard reaction. HMF is carcinogenic. However, the concentration does not exceed the limit recommended by Codex Alimentarius. During storage, the palm sugar will become darker, resulting from the Maillard reaction, especially if the sugar is stored in an environment with higher relative humidity and a high temperature. Hence, the sugar's cool, dry storage condition is recommended (Ho et al. 2007; Meenune and Naknean 2013; Naknean and Meenune 2015).

Similar to candy making in Western culture, the boiling of syrup starts at a lower boiling point, just slightly higher than 100°C. As more water evaporates and the sugar concentration becomes higher, the boiling point gradually rises. The physicochemical properties of the palm sap sugars vary greatly because most are from small production by farmers, whose techniques are different. Therefore, a vacuum evaporator is recommended for industrial-scale production as the sugar can be concentrated at lower temperatures and the browning reduced from Maillard and caramelization reactions (Naknean, Meenune, and Roudaut 2013).

Coconut palm sugar production is listed as part of the Thai National Heritage by UNESCO. The coconut tree is a multipurpose plant, and its fruit can be consumed as young coconut juice, or mature ones can be processed into coconut milk, as detailed in the previous section. Making coconut palm sugar is similar to the palmyra palm sugar process but is more commercially available. Sap from clusters of coconut flowers is collected mostly twice a day, and the concentration of sap into syrup or sugar solid is typically at a temperature between 100 and 120°C. Like palmyra palm sugar, coconut palm sugar has sweet aroma notes of caramel and creaminess, followed by smokiness. Other aromas detected and identified are pandan, vanilla, alcohol, acidic, roast, and nuttiness (Kabir and Lorjaroenphon 2014).

The brown color with its unique sweet aroma is used in desserts and is also crucial in many savory dishes. The sweetness from palm sugar adds dimension to food and helps to harmonize tastes and flavors from different ingredients into a well-balanced, delicate, and delicious recipe. Both palmyra and coconut palm sugars contain several volatile compounds yielding sweet, butter, and caramel aromas and include some acids, mineral salts, and proteins. This combination adds some sourness and saltiness and enhances more possibilities of reactions (due to amino acid content) compared to regular table sugar.

Table sugar is mainly from an industrialized process of refining sugar cane syrup into crystallized sucrose, while palm sugar is mostly from small-scale production. Though sucrose is also the primary component in palm sugar, table sugar is pure sucrose with no other compounds or elements. The sweetness of table sugar is often described as a sharp, flashy, isolated kind of sweetness as it merely delivers a load of sweet sensations on the tongue then disappears. Hence, light-colored foods and desserts require table sugar to maintain a desirable light color, e.g., sweetened sticky rice with coconut cream. For curry and other savory dishes, palm sugar will add layers of flavor to the food. However, the maximum amount added may be limited by the intensity of the brown color that is acceptable. For example, palm sugar complements the color and flavor of massaman curry very well because of its complexity (about 30 ingredients needed), dark color, and the right balance of salty, sour, and sweet tastes. Currently, there are many forms of palm sugar available, such as coconut flour syrup or crystallized palm sugar.

4.8 Lime Juice

Lime (*Citrus aurantifolia*) juice is ubiquitous in Thai cuisine. The lime fruit is green and smaller than lemon (*Citrus limon* L.). Their similarity is in the sourness provided by the citric acid in the juice, and both contain almost 50 g of citric acid per liter of juice (Penniston et al. 2008). Their aromas are similar to several identical compounds, but many compounds are found only in lime, especially those in the group of esters and oxides (Moshonas and Shaw 1972). In Thailand, only lime is used in Thai cuisine. Out of season, lime fruits are expensive. But, there are substitutes, including sour green mango and sour tamarind paste. Still, replacement by these is easily detectable and not always preferred by the consumer.

Thai cuisine incorporates contrasts of flavor, as in the famous *tom-yum*. The essential ingredients are herbs, such as lemongrass and kaffir lime leaves, for aroma, chili for spicy heat, and fish sauce for saltiness and sourness from lime juice. The combination of these pungency, saltiness, and sourness from three essential ingredients is almost like a fundamental pillar for many dishes. Hence, the ***nam-pla-prig*** (น้ำปลา พริก) or any dip usually contains a combination of these ingredients. The addition of lime juice gives a citrusy scent, which adds freshness to any dish. However, these aroma compounds are not very stable and sometimes do not last long enough for separation and identification in the laboratory (Chisholm, Wilson, and Gaskey 2003).

In many Thai dishes, lime juice has to be added last. Typically, the lime fruit is cut, and the seed is removed, then served on the side. Noodle soup and *tom-yum* are usually served in this manner to preserve the fresh citrus scent. In addition, papaya salad is sometimes served with lime with skin to ensure the citrus scent is not lost before reaching the consumer.

Lime juice provides the sourness and characteristic lime aroma of essential oil. The essential oil is located in the rind of the fruit. Thus the action of squeezing the lime helps to release the flavor compounds from the fruit. The aroma compound that has a significant role in the flavor of lime is citral, which is also found in lemon and lemongrass. Citral has the molecular structure of a terpene aldehyde and can be synthesized into a nature-identical flavor (to be used as a chemical food additive). Since the essential oil is very volatile, the flavor disappears quickly through evaporation. Thus, freshly squeezed lime juice has a better flavor profile and intensity than aged lime juice.

Food seasoned with freshly squeezed lime juice also taste better as there is no bitterness associated with aged lime juice. The fruit cell structure is damaged as the lime wedge is pressed to release the juice sac. The natural enzyme (limonin D-ring lactone hydrolase) is also released and then reacts with another terpene substance in lime, limonoids (limonoate A-ring lactone), resulting in the conversion of limonoids into bitter limonin. The process is called *delayed bitterness* (Wong 2018b). Whether for soup or spicy salad, lime is, thus, the last ingredient to be added to

Figure 4.5   Sour ingredients. (a) Lime fruit. (b) Tamarind fruit, pulp, and paste. (c) Sour green mango. (d) (clockwise) *Ma-aueg, ta-ling-pling, ma-dun.*

maximize the desirable citrusy aroma and prevent the development of a bitter taste. To enhance lime flavor, the whole cut lime (with the peel on but no seeds) is sometimes directly added to dishes such as spicy papaya salad (*som-tum*) as an ingredient itself. The practice is similar to how lemon tea is served in the West, i.e., with a lemon wedge on the side. Sourness can be obtained from other fruits, such as sour mango and tamarind (Fig 4.5 a-d). However, the aroma and color of finished dishes will vary by the characteristics of the replacement fruit.

4.9 Garlic

Garlic (*Allium sativum*) has been consumed in many regions in the world, especially Egypt, China and India, for over four centuries. It can be used in many forms (Fig. 4.6 a-d). Raw garlic can be sliced and used to flavor several spicy salads or *yum* and in condiments such as fish sauce and chili or *nam-pla-prig* (น้ำปลาพริก). Garlic has a pungent aroma due to its volatile sulfur compound thiosulfates such as allicin (Agarwal 1996; Macpherson et al. 2005). Garlic can be fermented by lactic acid bacteria to be pickled garlic (*gra-tiam-dong*, กระเทียมดอง) (Klayraung et al. 2008). Its use is for both sour and sweet tastes and fermented aromas. Pickled

Figure 4.6   (a) Thai garlic. (b) Imported garlic. (c) Shallot. (d) Pickled garlic.

garlic has a milder smell, and is not pungent or sharp like the fresh one. It goes with Chinese-style meat stew, e.g., chicken stew with pickled garlic, spicy salad, and various other food products.

Both garlic and onion are botanically located in the genus *Allium*. Garlic is *A. sativum,* and onion is *A. cepa.* The cells of Allium contain several sulfur-containing chemicals stored in the cytoplasm of the bulb cells. These sulfur compounds are odorless until the cells are damaged, and the compounds can be converted to aromatic and pungent volatiles, giving the characteristic odor of garlic and onion. Garlic and onion grown in sulfur-rich soil tend to contain more of the sulfur compounds and thus produce more robust aromas, pungency, and irritation to the eyes of the person who is chopping the onion.

The process starts when we damage the cloves (such as crushing garlic), thereby releasing an enzyme called alliinase that is initially stored in the vacuole part of the cell. The enzyme alliinase released from the vacuoles then reacts with sulfur compounds to produce pyruvate, ammonia, and a thiosulfinate. The thiosulfinate undergoes further reactions resulting in volatile compounds that are characteristic of onion and garlic. Allicin is a type of thiosulfinates that has been studied for its health benefits and is used as a quality indicator of garlic products. It also gives pungent and flavor to garlic (Jones et al. 2004; Ramirez et al. 2017).

As the action of enzyme alliinase is needed to develop the flavor substance "allicin," a cooking technique that causes more cells damages will produce higher flavor intensity. Thus, pounding garlic with a pestle in a mortar will give out a more potent garlic aroma than chopping with a sharp knife. Just as the browning of apples takes time to develop (after the enzymatic reaction of polyphenol oxidase and polyphenols), damaged garlic cells take about a minute for the full flavor and pungency from allicin to form. In addition, since most of the aromatic compounds in garlic are hydrophobic, cooking garlic in oil will produce a much more pungent odor than cooking in water (Joachim, Schloss, and Handel 2008).

For stir-fry dishes, a typical method is to crush unpeeled garlic cloves with the flat side of a knife, preferably a cleaver, then mince it further to the desired particle size. At this step, the thick garlic skin can be removed, while the thinner membrane may be left. A small amount of garlic skin gives a distinctive aroma when heated with oil. A temperature above 120°C produces flavor. Allicin in stir-fried garlic changes to diallyl trisulfide and diallyl disulfide (Kim et al. 1995).

The fried garlic can also be served on its own as a condiment with soupy dishes such as noodle soup and rice soup *(khao-tom,* ข้าวต้ม*)*. Garlic served as fried-garlic-with-oil *(kra-tiam-jeow,* กระเทียมเจียว*)* used as seasoning needs to be heated until it is slightly turning brown; there is a burnt aroma associated with this condiment.

Several studies have shown that different cooking methods, such as deep-oil frying, microwave-heating, oven-baking, roasting, and stir-frying, provide other volatile compounds. Dry heat provides more diallyl disulfide and diallyl trisulfide. For example, when garlic is heated with oil, the aroma comes from diallyl disulfide, methyl allyl disulfide, and vinyldithiins (Yu, Wu, and Ho 1993). Volatile compounds commonly associated with stir-frying are dimethyl sulfide, allyl alcohol, diallyl sulfide, methyl allyl disulfide, and diallyl disulfide (Kim et al. 1995).

Garlic has been known for its therapeutic properties in many food cultures, including Thailand. It has been used as a medicine in Egypt, India and China for centuries. With the advance in science, garlic has been proven repeatedly in many reports to have protective effects against degenerative diseases such as cardiovascular diseases, cancer, and immune dysfunction. Several other diseases involved with inflammation also benefit from garlic intake because garlic has several components with very high antioxidant activities. Since atherosclerosis or plaque that forms in the blood vessels, especially the coronary arteries, is a result of oxidation of blood cholesterol, LDL, the antioxidative properties of garlic help to reduce oxidative stress, thereby preventing the onset of atherosclerosis and reducing the risk of cardiovascular disease. Besides its antioxidation activities, the consumption of just half a clove of garlic has been shown to reduce blood cholesterol by as much as 9% (Santhosha, Jamuna, and Prabhavathi 2013).

Apart from their distinctive aroma, garlic and onion have several proven health benefits, such as antiplatelet activity that has a protective effect against cardiovascular diseases. In other words, the consumption of these compounds may reduce the risk of heart attack and stroke. The allicin and thiosulfinates in raw garlic contribute to this heart health benefit. Moderate cooking, such as baking at 200°C or boiling in water for 3 minutes, does not reduce the benefit. However, the heart health benefits of allicin will be lost after heating for more than 6 minutes. In order to prevent this loss, garlic should be crushed before cooking. The crushing allows allicin to react with the precursor to yield heat-stable organosulfur compounds. Since these compounds have the desirable antiplatelet activity, it is recommended to crush garlic before cooking (Cavagnaro et al. 2007).

4.10 Basil Leaves

There are many types of basil leaves used in Thai cuisine (Fig. 4.7 a-d). The type known as Thai basil or sweet basil (*Ocimum basilicum* L.) is known to Thais as *bai-ho-ra-pa* (ใบโหระพา). The most popular basil is holy basil (*Ocimum sanctum*) or *bai-ka-prao* (ใบกะเพรา). The spicy stir-fried meat dish with holy basil, eaten with crispy fried egg and rice, is well recognized as the most popular lunch item among Thais. The dish is called *pad-ka-prao* (ผัดกะเพรา), and it goes well with any type of meat (chicken, beef, pork) and seafood (shrimp). The most common holy basil has light green color throughout the plant; it is known as white holy basil (*ka-prao-khao*, กะเพราขาว). The other type has darker color stems and red-hue leaves; it is called red holy basil (*ka-prao-daeng*,

Figure 4.7   (a) White holy basil. (b) Red holy basil. (c) Sweet basil. (d) Mint.

กะเพราแดง). Red holy basil is not as widely commercially available. Although it provides a more intense holy basil characteristic aroma, the dark color is unfamiliar and deemed unappetizing to many. The scientific name for the white and red holy basil is *O. sanctum* L. cv. Khao and *O. sanctum* L. cv. Daeng, respectively (Raksakantong et al. 2011).

Any stir-fry dish starts with heating oil and crushed garlic in a wok; in the case of spicy stir-fried minced meat with holy basil (*pad-ka-prao*), chili is crushed and added to garlic for a hot and spicy sensation. The minced meat is added when a strong, pungent aroma has developed (it is very common to cause sneezing at this step). Finally, the basil is added after everything is already cooked, stirred very fast, and served right away. This common practice can be picked up easily even by an inexperienced cook because the heat-induced changes, i.e., the loss of desirable aroma and the darkening of cooked leaves, are easily detectable. A longer cooking time is likely to cause the leaves to turn dark, wilt, and unappealing overall. These characteristics are almost an indicator that the aroma of that dish has been altered. The *pad-ka-prao* dish is so popular that it has been credited as a culture-specific product representing Thai culture, similar to how kimchi represents Korean and matcha represents Japanese culture (Lee and Lopetcharat 2017).

The dominant aroma compound in holy basil leaves is eugenol. The odor description of eugenol is spicy and clove-like. Eugenol is also found in cloves, nutmeg, cinnamon, and bay leaves. Other odorants found in both fresh and cooked basil leaves are, for example, borneol that gives the soil a camphorous odor, 1,8-cineol, which smells minty, eucalyptus, and linalool with a floral, honeysuckle aroma. A compound with floral, applesauce scent characteristics, beta-damascenone, increases after cooking. In addition, some compounds emerge through the Maillard reaction induced by high heat. For example, 2,3-butanedione and 2-acetyl-pyrroline yield buttery, popcorn, and baked odors. However, these aromas can be mixed with some compounds (the structure is not identified) with strong skunky, pungent, catty, and green odors (Pojjanapimol, Chaiseri, and Cadwallader 2004). Thus, holy basil leaves should be heated only briefly to avoid the undesirable aroma compounds associated with cooked basil. It should be noted that, besides varieties such as white or red, the season, location, and stages of development affect the chemical composition and odor of basil leaves (Vani, Cheng, and Chuah 2009).

Some herbs are added just for their aromas, such as lemongrass and kaffir lime leaves. The herbs are left uneaten in the cookware. To release flavor, these two herbs need to be bruised or pounded, or torn slightly. Though not as much studied as is the case of garlic, damage to the plant cells before cooking may contribute to more enzyme activity that will not be able to emerge from the vacuole to react with other compounds (flavor precursors). Mint, chive, and cilantro are meant to be eaten but usually in small amounts since their other important purpose is dish decoration.

4.11 Herbs and Spices

Herbs and spices have characteristic aromas and flavors that are unique to them. Humans have consumed herbs and spices as food ingredients for over 6,000 years. In general, herbs refer to products derived from the leaves or flowers of plants. Spices are likely to come from the roots, bark, or seeds. Herbs and spices are not only used to add flavor to food but also they can be used as food additives for their antioxidant and antimicrobial activities. Some herbs and spices have additional health benefits, so they are commercially produced as food supplements, e.g., garlic and turmeric.

Figure 4.8   (a) Kaffir lime fruit and leaves. (b) Young peppercorns. (c) Ginger. (d) Turmeric.

Some herbs such as kaffir lime, fresh green peppercorn, ginger and turmeric are shown in Figures 4.8 a-d. Common herbs and spices in Thai cooking are also listed in Chapter 3, Tables 3.1 A-H. More details on the sensory properties of herbs and spices are in topics 3.3 *Aroma* and 3.4 *Chemesthesis*. In addition, details on cooking techniques of herbs and spices are in Chapter 5, topic 5.7 *Curry*.

There are two major groups of aromatic compounds in herbs and spices: phenolics and terpenes. Phenolic compounds are those that contain benzene rings with at least one hydroxyl group. Terpenes refer to molecules with a skeleton of at least five carbons in a zigzag formation. Terpene compounds are very volatile and are subject to cooking loss. Thus, their aromas are usually associated with the freshness of food (Zhuohong, Finley, and deMan 2018).

Aroma compounds from herbs and spices are primarily in the form of essential oils and are thus not soluble in water. Since plant cells are mostly water, these compounds need to be stored separately in various parts of plants that are in the oil glands. For example, citral, the major flavor compound in the essential oil of lemongrass, is accumulated in parenchyma oil cells or oil glands in the leaves (Croteau 2001; Lewinsohn et al. 1998). Herbs and spices with a hard texture, such as those rhizomes or dry seeds, are better with a long cooking time and mechanical action such as mincing or crushing in the preparation stage.

These actions will make the extraction of essential oils from the oil glands more efficient.

Besides giving aromas, many of these compounds are considered chemesthesis. They can stimulate receptors on the skin and create sensations such as burning, heating, tingling, warming, pungency, and penetrating anise. Aromatic compounds are mostly volatile, but the chemesthetic agents usually are not volatile. Thus, dried spices tend to have a higher proportion of chemesthesis than flavor because the volatile compounds have gone. The flavors and sensations can be perceived at a very low concentration. Thus, they are an addition at no expense to the other main ingredients (Loss and Bouzari 2016).

Each food culture has its own spice mixtures or pastes in its cuisine. Interestingly, the identity of Thai food, among over 13,000 recipes from several food cultures, is based on garlic and galangal (Table 3.1C). More than half of Thai food recipes contain garlic, but garlic is used frequently in other food cultures such as Central and South American as well. Although galangal is used in only about 10% of the Thai food recipes, it almost does not exist in other food cultures (Kopf 2015; Rush 2015).

On a smaller scale, just within Southeast Asian countries, coriander roots, seeds, and leaves were identified as unique to Thai cuisine (Panyayong 2016). Only a few menu items call for a substantial amount of coriander leaves, e.g., sago ball with minced pork filling (*sa-koo-sai-moo*, สาคูไส้หมู). Mostly, coriander leaves are used as a garnish at the top of any dishes or appetizers (similar to Hors d'oeuvres) simply to elevate the appearance of the food, so much so that the practice (*pug-chee-roi-nha*, ผักชีโรยหน้า) has become a phrase that describes something more attractive than what it actually is (in other words, window dressing). Conversely, coriander root is invisible, so most people are unaware of its presence in their food. The root is rarely used by itself but is almost always grounded together with garlic and pepper. Hence, the grounded mixture of coriander root, garlic, and pepper is nicknamed "three-friends" (*saam-gler*, สามเกลอ). The mixture can be used in a soup, meat margination, and various other dishes (see example on topic 7.10, spicy mixed salad recipe).

Thai food was also influenced by Indian food tradition but with some adaptations. For example, Indian garam masala is a mixture of mostly dried spices, i.e., cumin, coriander, cardamom, black pepper, cloves, mace, and cinnamon. On the other hand, Thai green curry paste is a combination of fresh herbs (green cayenne pepper, green bird's eye chili, galangal, lemongrass, kaffir lime zest, coriander root, red turmeric, red shallot, and garlic) and dried spices (white peppercorns, coriander seeds, cumin seeds). David Thompson (2002), an Australian chef and the author of a Thai Food cookbook, describes the use of herbs and spices in Thai curry paste as "dried spices should be used gingerly in Thai curries. They are there merely to add a subtle piquancy, to support and give dimension to the taste and fragrance."

Possibly, with the addition of fresh herbs, both in the paste and at a later heating stage, Thai curry clearly distinguishes its character from other cuisines' spice mixtures. Note that in Thai green curry paste, the only non-plant ingredient is fermented shrimp (ga-pi). More details on the functions of ferment shrimp in Thai food are detailed in the umami topic (Chapter 3).

The flavor and chemesthetic compounds in herbs and spices are mostly fat-soluble. A curry that contains coconut-milk-based curry is thus more aromatic. The traditional method of making Thai curry requires coconut cream emulsion to break down before adding curry paste. The technique involves heating the coconut cream until the oil layer is visibly separated from the milky constituent. Adding curry at this point releases intense aromas as the oil acts as a solvent to extract the fat-soluble flavors and chemesthesis.

"Jungle curry" (gaeng-pa, แกงป่า) is a type of curry with no coconut milk. The name implies that the curry is cooked and eaten during a journey away from home. Travelers can conveniently bring curry paste along but not the highly perishable coconut milk. Because there is no lipid (fat or oil) ingredient, the extraction of the flavor and chemesthetic compounds from the curry paste is not efficient. Thus, the sensory attributes of water-based curry are not comparable to the coconut-milk-based curry.

4.12 Limestone Water

Each cell of any fruits and vegetables is surrounded by a cell wall. Water kept inside the cell wall contributes to the freshness of fruits and vegetables, while the structure of the cell wall influences firmness. The cell wall and the middle lamella contain a polymer called pectin, which is a type of non-starch polysaccharides. When the fruit is not yet fully ripe, pectin is very strong and holds cell walls together tightly. As the fruit continues to ripe, the pectin is weakened, leading to the softness of the fruit. The addition of calcium ion (Ca^{2+}) by dipping or immersion of fruits or vegetable pieces in a solution of calcium salts help strengthen the pectin by creating crosslink or formation of bridges between chains of pectin. The calcium salt use includes calcium lactate, calcium gluconate, calcium chloride, and calcium hydroxide (Alonso, Canet, and Rodriguez 1997; Lovera, Ramallo, and Salvadori 2014; Odake, Otoguro, and Kaneko 1999). The firming effect, however, does not apply to every fruit. While mango, tomato, ume (Japanese plum), and cantaloupe gain firmness with calcium chloride dip, the apple and pineapple do not (Sirijariyawat and Charoenrein 2014; Trisnawati, Aurum, and Sugianyar 2019).

In Thai cuisine, fruit cuts, such as green mango, tamarind, and wax gourd, are soaked in limestone water until the texture is firm (usually 1-3 hours), washed away the limestone water, and are ready to be pickled. Pickling is generally sweet, so the firm fruit cuts are to be submerged in the mixture of sugar and salt. The sweet mango pickle is "*chair-im*" (มะม่วงแช่อิ่ม). The more sour one is "*dong*" (มะม่วงดอง). The tart fruit pickle is usually served with dried chili and salt dipping powder (*prig-gub-gluar*, พริกกับเกลือ). Pumpkin pieces are soaked in the limestone water before boiling with coconut milk and sugar to make a pumpkin-coconut cream desert (*fug-tong-gaeng-buad*, ฟักทองแกงบวด). Without the calcium ion that helps strengthen the cell wall, the fruit or vegetable would be soft, mushy, and cannot hold its shape. Baking soda (sodium bicarbonate, $NaHCO_3$) cannot replace limestone water for this function because the needed active ingredient is the calcium ion, which is not available from the baking soda.

The alkalinity of the limestone water is required in some products for its functions, such as, providing a firmer and elastic texture and contributing to unique flavor attributes (Mojarrad and Rafe 2018; Qin et al. 2019). For example, the sweetened starch paste (*kha-nhom-piag-poon*, ขนมเปียกปูน) and crispy deep-fried batters such as *kha-nhom-dog-jog* (ขนมดอกจอก) and *kluay-tod* (กล้วยทอด).

Calcium hydroxide is produced from high-temperature (700–900°C) burning of limestone or shells, a process dated back to the sixth century in the Dvaravati Period (as early as the seventh century). After this heating step, the calcium carbonate ($CaCO_3$) will turn into calcium oxide (CaO), and the product is called "quick lime" (*poon-dib*, ปูนดิบ). When adding water (slaking) to the quick lime, the calcium hydroxide is obtained, and the product is now called "slaked lime" (*poon-sook*, ปูนสุก). With high moisture, the consistency of the slaked lime is like that of a starch paste (Wongsawan 2002). Traditionally, turmeric can be added to this paste to obtain red color, and this product is called *poon-daeng* (ปูนแดง) indicating the red color of the limestone solution. Note that the curcumin pigment in turmeric changes from yellow to red hue at pH 7.5–8.5 (Priyadarsini 2014).

Before using the limestone solution in the recipe, simply add water to the limestone paste, wait until all the particles settle to the bottom, and scoop only the clear solution for use. Since the liquid is clear and free of color, this ingredient is the limestone water (*nam-poon-sai*, น้ำปูนใส).

4.13 Conclusion

The culinary tradition of savory dishes featured fermented plants or aquatic animals as a foundation for umami taste, regardless of whether the finished dish is vegetables or meat-based. Then, the generous use of herbs and spices added complexity and identity to each dish. Though garlic is used most frequently, galangal is the most distinctive ingredient hardly found in other cuisines. Thai people are particular about the sensory properties of the food. Each element can vary from breed, plant season, maturity, and storage time. Scientific research offers insights into the physicochemical properties of ingredients and supports why chefs, restaurants, or just about any Thai person have a specific preference for ingredient sourcing.

References

Agarwal, Kailash C. 1996. "Therapeutic actions of garlic constituents." *Medicinal Research Reviews* 16 (1):111–124.

Ajinomoto Academic Department. 2011. "Umami as source of deliciousness (อู มา มิ ที่ มา ของ ความ อร่อย)." *Journal of Food Technology, Siam University* 6 (1):17–26.

Alonso, Jesus, Wenceslao Canet, and Teresa Rodriguez. 1997. "Thermal and calcium pretreatment affects texture, pectinesterase and pectic substances of frozen sweet cherries." *Journal of Food Science* 62 (3):511–515.

Ariyaprakai, Suwimon, Tanachote Limpachoti, and Pasawadee Pradipasena. 2013. "Interfacial and emulsifying properties of sucrose ester in coconut milk emulsions in comparison with Tween." *Food Hydrocolloids* 30 (1):358–367.

Aza-González, Cesar, Hector G. Núñez-Palenius, and Neftalí Ochoa-Alejo. 2011. "Molecular biology of capsaicinoid biosynthesis in chili pepper (*Capsicum* spp.)." *Plant Cell Reports* 30 (5):695–706.

Bayarri, Sara, Andrew J. Taylor, and Joanne Hort. 2006. "The role of fat in flavor perception: effect of partition and viscosity in model emulsions." *Journal of Agricultural and Food Chemistry* 54 (23):8862–8868.

Cavagnaro, Pablo F., Alejandra Camargo, Claudio R. Galmarini, and Philipp W. Simon. 2007. "Effect of cooking on garlic (*Allium sativum* L.) antiplatelet activity and thiosulfinates content." *Journal of Agricultural and Food Chemistry* 55 (4):1280–1288.

Chiewchan, Naphaporn, Chanthima Phungamngoen, and Suwit Siriwattanayothin. 2006. "Effect of homogenizing pressure and sterilizing condition on quality of canned high fat coconut milk." *Journal of Food Engineering* 73 (1):38–44.

Chisholm, Mary G., Matthew A. Wilson, and Gina M. Gaskey. 2003. "Characterization of aroma volatiles in key lime essential oils (*Citrus aurantifolia* Swingle)." *Flavour and Fragrance Journal* 18 (2):106–115.

Cholamarkpijarn, Mom Luang Tiew. 1980. "Vegetable Groups." In *Tumrub Sai Yaowapa*, edited by Yaovabha Bongsanid, 51–60. Bangkok, Thailand: Saipuna Samakom.

Choorit, Wanna., and Poonsuk. Prasertsan. 1992. "Characterization of proteases produced by newly isolated and identified proteolytic microorganisms from fermented fish (Budu)." *World Journal of Microbiology and Biotechnology* 8 (3):284–286. doi: 10.1007/BF01201880

Chotechuang, Natida. 2012. "Taste active components in Thai foods: a review of Thai traditional seasonings." *Journal of Nutrition & Food Sciences* S10:004. doi:10.4172/2155-9600.S10-004

Chukeatirote, Ekachai. 2015. "Thua nao: Thai fermented soybean." *Journal of Ethnic Foods* 2 (3):115–118. doi: 10.1016/j.jef.2015.08.004

Croteau, Rodney. 2001. "Oil-filled glands on a leaf." *Trends in Plant Science* 6 (9):439.

Dajanta, Katekan, Arunee Apichartsrangkoon, Ekachai Chukeatirote, and Richard A. Frazier. 2011. "Free-amino acid profiles of thua nao, a Thai fermented soybean." *Food Chemistry* 125 (2):342–347. doi: 10.1016/j.foodchem.2010.09.002

Daroonpunt, Rungsima, Somboon Tanasupawat, Takuji Kudo, Moriya Ohkuma, and Takashi Itoh. 2016. "*Virgibacillus kapii* sp. nov., isolated from Thai shrimp paste (Ka-pi)." *International Journal of Systematic and Evolutionary Microbiology* 66 (4):1832–1837. doi: 10.1099/ijsem.0.000951

Eggleston, Gillian., John W. Finley, and John. M. deMan. 2018. "Carbohydrates." In *Principles of Food Chemistry*. John M. deManJohn W. FinleyW. Jeffrey HurstChang Yong Lee, Switzerland: Springer.

Foreign Office of the Government Public Relations Department Thailand. n/d. *Thailand the Kitchen of the World*. Thailand: The Government Public Relations Department.

Foster-Powell, Kaye, Susanna H. A. Holt, and Janette C. Brand-Miller. 2002. "International table of glycemic index and glycemic load values: 2002." *The American Journal of Clinical Nutrition* 76 (1):5–56.

Giacometti, T. 1979. "Free and bound glutamate in natural products." In *L. J. Filer, S. Garattini, M. R. Kare, W. A. Reynolds, & R. J. Wurtman (Eds.), Glutamic acid: Advances in biochemistry and physioology (pp. 25–34). New York: Raven: Advances in Biochemistry and Physiology*, 25–34.

Hajeb, Parvaneh., and Selamat. Jinap. 2015. "Umami taste components and their sources in Asian foods." *Critical Reviews in Food Science and Nutrition* 55 (6):778–791.

Ho, Chunwei. W., W. Mustafa. Wan Aida, Mohamad Yusuf Maskat, and Hasna. Osman. 2007. "Changes in volatile compounds of palm sap (*Arenga pinnata*) during the heating process for production of palm sugar." *Food Chemistry* 102 (4):1156–1162.

Joachim, David, Andrew Schloss, and Philip A. Handel. 2008. "Allium." In *The Science of Good Food: The Ultimate Reference on How Cooking Works*, David Joachim, Andrew Schloss, 19–22. Canada: Robert Rose Inc.

Jones, Meriel G., Jill Hughes, Angela Tregova, Jonothan Milne, A. Brian Tomsett, and Hamish A. Collin. 2004. "Biosynthesis of the flavour precursors of onion and garlic." *Journal of Experimental Botany* 55 (404):1903–1918.

Kabir, Alamgir, and Yaowapa Lorjaroenphon. 2014. "Identification of aroma compounds in coconut sugar." Proceedings of 52nd Kasetsart University Annual Conference: Agro-Industry. Kasetsart University, Thailand.

Kim, Sun Min, Chung May Wu, Akio Kobayashi, Kikue Kubota, and Joji Okumura. 1995. "Volatile compounds in stir-fried garlic." *Journal of Agricultural and Food Chemistry* 43 (11):2951–2955.

Klayraung, Srikanjana, Helmut Viernstein, Jakkapan Sirithunyalug, and Siriporn Okonogi. 2008. "Probiotic properties of Lactobacilli isolated from Thai traditional food." *Scientia Pharmaceutica* 76 (3):485–504.

Kongpan, Srisamorn. 2018. *Intangible Cultural Heritage Foods of Thailand* (อาหารขึ้นทะเบียน มรดกทางภูมิปัญญาของชาติ) Bangkok, Thailand: S.S.S.S. (บริษัท ส.ส.ส.ส. จำกัด).

Kopf, Dan. 2015. "What Are the Defining Ingredients of a Culture's Cuisine?." *Priceonomics*, accessed June 23. https://priceonomics.com/what-are-the-defining-ingredients-of-a-cultures/.

Kraikruan, Wilawan, Sutevee Sukprakarn, Orarat Mongkolporn, and Sirikul Wasee. 2008. "Capsaicin and dihydrocapsaicin contents of Thai chili cultivars." *Kasetsart Journal (Natural Science)* 42:611–616.

Kumar, Rajesh, Neeraj Dwivedi, Rakesh Kumar Singh, Sanjay Kumar, Ved Prakash Rai, and Major Singh. 2011. "A review on molecular characterization of pepper for capsaicin and oleoresin." *International Journal of Plant Breeding and Genetics* 5:99–110.

Kwanbunjan, Karunee. 2020. *Diseases of the Modern World: The Role of Nutrition, Lifestyle and Genetics*, 1st ed. Bangkok, Thailand: Hua Nam Printing Co., Ltd.

Kwon, Kisung, Kwan Hwa Park, and Khee Choon Rhee. 1996. "Fractionation and characterization of proteins from coconut (*Cocos nucifera* L.)." *Journal of Agricultural and Food Chemistry* 44 (7):1741–1745.

Lapsongphon, Nawaporn, Jirawat Yongsawatdigul, and Keith R. Cadwallader. 2015. "Identification and characterization of the aroma-impact components of Thai fish sauce." *Journal of Agricultural and Food Chemistry* 63 (10):2628–2638.

Lee, Hye-Seong, and Kannapon Lopetcharat. 2017. "Effect of culture on sensory and consumer research: Asian perspectives." *Current Opinion in Food Science* 15:22–29. doi: https://doi.org/10.1016/j.cofs.2017.04.003

Lewinsohn, Efraim, Nativ Dudai, Yaakov Tadmor, Irena Katzir, Uzi Ravid, Eli Putievsky, and Daniel M. Joel. 1998. "Histochemical localization of citral accumulation in lemongrass leaves (*Cymbopogon citratus* (DC.) Stapf., Poaceae)." *Annals of Botany* 81 (1):35–39.

Li, Hongyan, Sangeeta Prakash, Timothy M. Nicholson, Melissa A. Fitzgerald, and Robert G. Gilbert. 2016. "The importance of amylose and amylopectin fine structure for textural properties of cooked rice grains." *Food Chemistry* 196:702–711.

Lioe, Hanifah Nuryani, Jinap Selamat, and Masaaki Yasuda. 2010. "Soy sauce and its umami taste: a link from the past to current situation." *Journal of Food Science* 75 (3):R71–R76. doi: 10.1111/j.1750-3841.2010.01529.x

Loss, Christopher R., and Ali Bouzari. 2016. "On food and chemesthesis – food science and culinary perspectives." In *Chemesthesis: Chemical Touch in Food and Eating*, 250. Shane T McDonald, David A. Bolliet, John E Hayes.

Lovera, Nancy, Laura Ramallo, and Viviana Salvadori. 2014. "Effect of processing conditions on calcium content, firmness, and color of papaya in syrup." *Journal of Food Processing* 2014. 1-8.

Macpherson, Lindsey J., Bernhard H. Geierstanger, Veena Viswanath, Michael Bandell, Samer R. Eid, SunWook Hwang, and Ardem Patapoutian. 2005. "The pungency of garlic: activation of TRPA1 and TRPV1 in response to allicin." *Current Biology* 15 (10):929–934.

Mojarrad, Lida, and Ali Rafe. 2018. "Effect of high-amylose corn starch addition on canning of yellow alkaline noodle composed of wheat flour and microbial transglutaminase: Optimization by RSM." *Food Science & Nutrition* 6. doi: 10.1002/fsn3.667.

Mao, Like, Yrjö H. Roos, Costas G. Biliaderis, and Song Miao. 2017. "Food emulsions as delivery systems for flavor compounds: a review." *Critical Reviews in Food Science and Nutrition* 57 (15):3173–3187.

Meenune, Mutita, and Phisut Naknean. 2013. "Moisture sorption isotherm and glass transition of palm sugar cake as affected by storage temperature." 2nd IPCBEE 53:13.

Ministry of Agriculture and Cooperatives, National Bureau of Agricultural Commodity and Food Standards. 2003. Thai Agricultural Standard TAS 4000-2003, Thai Hom Mali Rice.

Moshonas, Manuel G., and Philip E. Shaw. 1972. "Analysis of flavor constituents from lemon and lime essence." *Journal of Agricultural and Food Chemistry* 20 (5):1029–1030.

Naknean, Phisut, and Mutita Meenune. 2015. "Impact of clarification of palm sap and processing method on the quality of palm sugar syrup (*Borassus flabellifer* Linn.)." *Sugar Tech* 17 (2):195–203.

Naknean, Phisut, Mutita Meenune, and Gaëlle Roudaut. 2010. Characterization of Palm Sap Harvested in Songkhla Province, Southern Thailand.

Naknean, Phisut, Mutita Meenune, and Gaëlle Roudaut. 2013. "Changes in properties of palm sugar syrup produced by an open pan and a vacuum evaporator during storage." *International Food Research Journal* 20 (5):2323.

National Bureau of Agricultural Commodity and Food Standards, Ministry of Agriculture and Cooperative. 2019. Food Consumption Data of Thailand. Thailand.

Odake, Sachiko, Chikao Otoguro, and Kentaro Kaneko. 1999. "Effect of calcium gluconate addition on the properties of ume fruit (Japanese apricot) products." *Food Science and Technology Research* 5 (2):227–233.

Panyayong, Chunkamol. 2016. "Study of ASEAN food culture for ASEAN soci-cultural community." *Institute of Culture and Arts Journal* 18 (1):43–52.

Parnsahkorn, Sunan, and Jutarong. Langkapin. 2013. "Changes in physico-chemical characteristics of germinated brown rice and brown rice during storage at various temperatures." *Agricultural Engineering International: CIGR Journal* 15 (2):293–303.

Patil, Umesh, and Soottawat Benjakul. 2017. "Characteristics of albumin and globulin from coconut meat and their role in emulsion stability without and with proteolysis." *Food Hydrocolloids* 69:220–228.

Patil, Umesh, and Soottawat Benjakul. 2018. "Coconut milk and coconut oil: their manufacture associated with protein functionality." *Journal of Food Science* 83 (8):2019–2027.

Penniston, Kristina L., Stephen Y. Nakada, Ross P. Holmes, and Dean G. Assimos. 2008. "Quantitative assessment of citric acid in lemon juice, lime juice, and commercially-available fruit juice products." *Journal of Endourology* 22 (3):567–570.

Peters, Michael, Kevin Mackenzie, and Pam Bryden. 2002. "Finger length and distal finger extent patterns in humans." *American Journal of Physical Anthropology: The Official Publication of the American Association of Physical Anthropologists* 117 (3):209–217.

Pojjanapimol, Sompoche, Siree Chaiseri, and Keith R. Cadwallader. 2004. *Heat-Induced Changes in Aroma Components of Holy Basil (Ocimum sanctum L.)*. New York, NY: Marcel Dekker.

Priyadarsini, Kavirayani Indira. 2014. "The chemistry of curcumin: from extraction to therapeutic agent." *Molecules* 19 (12):20091–20112.

Qin, Yang, Hui Zhang, Yangyong Dai, Hanxue Hou, and Haizhou Dong. 2019. "Effect of alkali treatment on structure and properties of high amylose corn starch film." *Materials* 12 (10):1705.

Raghavendra, S. N., and K. S. M. S. Raghavarao. 2010. "Effect of different treatments for the destabilization of coconut milk emulsion." *Journal of Food Engineering* 97 (3):341–347.

Raksakantong, Pornpimol, Sirithon Siriamornpun, Jiranan Ratseewo, and Naret Meeso. 2011. "Optimized drying of kaprow leaves for industrial production of holy basil spice powder." *Drying Technology* 29 (8):974–983. doi: 10.1080/07373937.2011.558649.

Ramesh, M., K. R. Bhattacharya, and J. R. Mitchell. 2000. "Developments in understanding the basis of cooked-rice texture." *Critical Reviews in Food Science and Nutrition* 40 (6):449–460.

Ramirez, Daniela, Daniela Locatelli, Roxana González, Pablo Cavagnaro, and Alejandra Camargo. 2017. "Analytical methods for bioactive sulfur compounds in Allium: an integrated review and future directions." *Journal of Food Composition and Analysis*. doi: 10.1016/j.jfca.2016.09.012

Relkin, Perla, Marjorie Fabre, and Elisabeth Guichard. 2004. "Effect of fat nature and aroma compound hydrophobicity on flavor release from complex food emulsions." *Journal of Agricultural and Food Chemistry* 52 (20):6257–6263.

Ritthiruangdej, Pitiporn, and Thongchai Suwonsichon. 2007. "Relationships between NIR spectra and sensory attributes of Thai commercial fish sauces." *Analytical Sciences* 23 (7):809–814.

Rotsatchakul, Premsiri, Siree Chaiseri, and Keith R. Cadwallader. 2007. "Identification of characteristic aroma components of Thai fried chili paste." *Journal of Agricultural and Food Chemistry* 56 (2):528–536.

Rush, James. 2015. "The Ingredients that Make National Dishes National Dishes." *The Independent*, Last Modified 22 July 2015, accessed 22 June. https://www.independent.co.uk/life-style/food-and-drink/news/national-cuisines-what-ingredients-make-dishes-from-different-cultures-distinctive-10404837.html.

Saisithi, Prasert. 1994. "Traditional fermented fish: fish sauce production." In *Fisheries Processing*, 111–131. A.M. Martin, Department of Biochemistry Memorial University of NewfoundlandSt John'sCanada Springer.

Santhosha, S. G., Prakash Jamuna, and S. N. Prabhavathi. 2013. "Bioactive components of garlic and their physiological role in health maintenance: a review." *Food Bioscience* 3:59–74.

Saputro, Arifin Dwi, Davy Van de Walle, and Koen Dewettinck. 2019. "Palm sap sugar: a review." *Sugar Tech* 1–6. 862-867

Seow, Chee C., and Choon N. Gwee. 1997. "Coconut milk: chemistry and technology." *International Journal of Food Science & Technology* 32 (3):189–201.

Sirijariyawat, Arpassorn, and Sanguansri Charoenrein. 2014. "Texture and pectin content of four frozen fruits treated with calcium." *Journal of Food Processing and Preservation* 38 (3):1346–1355.

Srikaeo, Khongsak., and Namphung Thongta. 2015. "Effects of sugarcane, palm sugar, coconut sugar and sorbitol on starch digestibility and physicochemical properties of wheat based foods." *International Food Research Journal* 22 (3).923-929

Srisajjalertwaja, Siriwan, Arunee Apichartsrangkoon, Pittaya Chaikham, Yasinee Chakrabandhu, Pattavara Pathomrungsiyounggul, Noppol Leksawasdi, Wissanee Supraditareporn, and Sathira Hirun. 2012. "Color, capsaicin and volatile components of baked Thai green chili (*Capsicum annuum* Linn. var. Jak Ka Pat)." *Journal of Agricultural Science* 4 (12):75.

Suriyanont, Atchariya, and Chuenjit Chancharoonpong. 2018. "Making process and factors associating to properties of fermented fish (Plara) in some parts of upper northeast area of Thailand." *Burapha Science Journal (วารสาร วิทยาศาสตร์ บูรพา)* 23 (1):566–578.

Tananuwong, Kanitha, and Sittiwat Lertsiri. 2010. "Changes in volatile aroma compounds of organic fragrant rice during storage under different conditions." *Journal of the Science of Food and Agriculture* 90 (10):1590–1596. doi: 10.1002/jsfa.3976

Tananuwong, Kanitha, and Yuwares Malila. 2011. "Changes in physicochemical properties of organic hulled rice during storage under different conditions." *Food Chemistry* 125 (1):179–185.

Tanasupawat, Somboon, Nitcha Chamroensaksri, Takuji Kudo, and Takashi Itoh. 2010. "Identification of moderately halophilic bacteria from Thai fermented fish (pla-ra) and proposal of *Virgibacillus siamensis* sp. nov." *The Journal of General and Applied Microbiology* 56 (5):369–379.

Tangsuphoom, Nattapol, and John N. Coupland. 2005. "Effect of heating and homogenization on the stability of coconut milk emulsions." *Journal of Food Science* 70 (8):e466–e470.

Tangsuphoom, Nattapol, and John N. Coupland. 2009. "Effect of thermal treatments on the properties of coconut milk emulsions prepared with surface-active stabilizers." *Food Hydrocolloids* 23 (7):1792–1800. doi: https://doi.org/10.1016/j.foodhyd.2008.12.001

The Rice Trader. 2019. "The 11th Annual – World's Best Rice Contest 2019." accessed December 3. https://thericetrader.com/conferences/2019-wrc-manila/worlds-best-rice/.

Thompson, David. 2002. *Thai Food*. London, UK: Pavillion Books.

Tinchan, Patcharaporn, Yaowapa Lorjaroenphon, Keith R. Cadwallader, and Siree Chaiseri. 2015. "Changes in the profile of volatiles of canned coconut milk during storage." *Journal of Food Science* 80 (1):C49–C54.

Trinidad, Trinidad P., Aida C. Mallillin, Rosario S. Sagum, and Rosario R. Encabo. 2010. "Glycemic index of commonly consumed carbohydrate foods in the Philippines." *Journal of Functional Foods* 2 (4):271–274. doi: https://doi.org/10.1016/j.jff.2010.10.002

Trisnawati, Wahju, Fawzan Sigma Aurum, and M. Sugianyar. 2019. "Application of Calcium Hydroxide Concentration and Immersion Duration Towards Tomato Sweets Quality." *IOP Conference Series: Earth and Environmental Science.*

Vanavichit, Apichart, Wintai Kamolsukyeunyong, Meechai Siangliw, Jonaliza L. Siangliw, Suniyom Traprab, Siriphat Ruengphayak, Ekawat Chaichoompu, Chatree Saensuk, Ekapol Phuvanartnarubal, Theerayut Toojinda, and Somvong Tragoonrung. 2018. "Thai Hom Mali Rice:

Origin and Breeding for Subsistence Rainfed Lowland Rice System." *Rice* 11 (1):20. doi: 10.1186/s12284-018-0212-7

Vani Krishna, S. Raseetha, Sin Fun Cheng, and C. H. Chuah. 2009. "Comparative study of volatile compounds from genus *Ocimum*." *American Journal of Applied Sciences* 6 (3):523.

Wattanapahu, Saowapark, Thongchai Suwonsichon, Wannee Jirapakkul, and Sumaporn Kasermsumran. 2012. "Categorization of coconut milk products by their sensory characteristics." *Agriculture and Natural Resources* 46 (6):944–954.

Weber, Steve, Heather Lehman, Timothy Barela, Sean Hawks, and David Harriman. 2010. "Rice or millets: early farming strategies in prehistoric central Thailand." *Archaeological and Anthropological Sciences* 2 (2):79–88. doi: 10.1007/s12520-010-0030-3

Wichaphon, Jetsada, Chaufah Thongthai, Apinya Assavanig, and Sittiwat Lertsiri. 2012. "Volatile aroma components of Thai fish sauce in relation to product categorization." *Flavour and Fragrance Journal* 27 (2):149–156.

Wong, Dominic W. S. 2018a. "Carbohydrates." In *Mechanism and Theory in Food Chemistry*, Dominic W.S. Wong, Springer, ISBN-10: 3319507656 2nd ed., 157–162. Springer.

Wong, Dominic W. S. 2018b. "Flavors." In *Mechanism and Theory in Food Chemistry*, Dominic W.S. Wong, Springer, ISBN-10: 3319507656 277–321. Springer.

Wongsawan, Sukkamol. 2002. *"Analysis of Lime from Clams Unearthed at Wang Phai Archaeological Site, Ban Mi District, Lop Buri Province."* Masters of Arts, Department of Archaeology, Silpakorn University, Thailand.

Yankowski, Andrea, Puangtip Kerdsap, and Nigel Chang. 2015. ""Please pass the salt" – an ethnoarchaeological study of salt and salt fermented fish production, use and trade in Northeast Thailand." *Journal of Indo-Pacific Archaeology* 37:4–13.

Yongsawatdigul, Jirawat., Surelak. Rodtong, and Nongnuch. Raksakulthai. 2007. "Acceleration of Thai fish sauce fermentation using proteinases and bacterial starter cultures." *Journal of Food Science* 72 (9):M382–M390.

Yu, Tung Hsi, Chung May Wu, and Chi Tang Ho. 1993. "Volatile compounds of deep-oil fried, microwave-heated and oven-baked garlic slices." *Journal of Agricultural and Food Chemistry* 41 (5):800–805.

Zhuohong, Xie, John W. Finley, and John M. deMan. 2018. In "Herbs and spices, *Principles of Food Chemistry*, ISBN: 978-3-319-63607-8, 457–481. Springer.

5

SCIENTIFIC PRINCIPLES IN THAI CUISINE COOKING TECHNIQUE

VALEERATANA K. SINSAWASDI AND NITHIYA RATTANAPANONE

Contents

5.1 Fundamentals of Thai Cuisine Cooking Techniques

Traditional cooking techniques were limited to cooking on medium heat with little or no oil, mostly grilling and boiling. Then, the Chinese introduced relatively new cooking methods such as steaming and stir-frying, and Westerners introduced baking. Eventually, with the availability of kitchen tools, the development of contemporary cooking techniques such as boiling with curry paste (*gaeng*, แกง), stir-frying (*pud*, ผัด), mixing with chili and lime (*yum*, ยำ), and deep-frying (*tod*, ทอด) were developed (Seubsman et al. 2009).

DOI: 10.1201/9781003182924-7

Dishes such as chili dip (*nam-prig*, น้ำพริก) and spicy salad (*yum*, ยำ) do not require heat to cook. However, most other dishes need heat to transform the raw materials to achieve the desired color, texture, taste, and aroma in the finished product. Heat is also required for non-sensory-related reasons, such as to ensure safety since it can destroy pathogenic and spoilage microorganisms that potentially cause sickness and food spoilage. Apart from sensory reasons, heat can also extend the shelf life of the food as it reduces the microbial load. Heat is a form of energy that causes the food molecules to increase their speed. As the molecules move and bump into each other, the food becomes hot, and several physicochemical changes occur. The starch in rice grains and other starchy foods become gelatinized and edible with moisture and heat. As for another component, protein is denatured with heat. For example, egg white is transformed from a clear liquid to become an opaque and white solid. The heat reaction with carbohydrate and protein creates a brown color and flavor from the Maillard reaction, for example, in grilled meat that turns brown and becomes aromatic. Solid fat such as lard is melted into the oil with heat, while an emulsion of oil in water, such as coconut milk, breaks into separate layers of oil and water (Mauer, Martin-Gonzalez, and Fernanda 2017).

Thai people have employed both dry-heat and moist-heat cooking methods. Moist heat, like boiling, is used for cooking soup and curries. Blanching and steaming are mainly used to prepare vegetables as a side dish. Boiling is also used to prepare many traditional desserts, mainly with a combination of coconut milk and palm sugars with varieties of add-ons such as sticky rice, taro, banana, and pumpkin. The most common dry-heat cooking technique used is grilling. Besides the grilling of several types of meat, many dishes require the preparation step of grilling raw materials, such as shallots and chili peppers. Roasting larger raw materials such as meat was less common as the oven was typically not a part of the Thai kitchen until just recently.

Tumrub Sai Yaowapa (ตำรับสายเยาวภา) is a significant cookbook that is lauded as the first modern Thai cookbook with modern systematic measurement units in each recipe. An astonishing 87 cooking terms are used to describe preparation and cooking methods and techniques (Bongsanid 1935). For example, terms for moist heat are, for example, *tom, hoong, luag, nueng, toon, ruan,* and *piak* (ต้ม, หุง, ลวก, นึ่ง, ตุ๋น, รวน,

เปียก). Terms used for dry-heat methods are, for example, *ping, yang, pao, larm, lhon,* and *kua* (ปิ้ง, ย่าง, เผา, หลาม, หลน, คั่ว). Mixing techniques are also very specific and precise about the expected results in each dish, for example, *ka-yum, ka-yee, tee, kon, kloog, klao,* and *guan* (ขยำ, ขยี้, ตี, คน, คลุก, เคล้า, กวน). Each term was carefully defined to distinguish actions that are, in fact, almost identical in the eyes of a layperson. For example, *kloog* means to mix several ingredients lightly until taste and flavor are homogenous, while *klao* requires a light press of food components for a softer texture. Each dish of Thai food is thus a culinary work of art and the result of years of cumulative experience and perfected techniques.

The science behind each ingredient and the cooking technique can be explained in lengthy detail, and the art of cooking can be practiced and passed down through family tradition or formal training. However, Thais believe that it is not simply knowledge or skill needed to achieve a flavorful, appealing, and delicious dish. In the same way, as Westerners use the term "green fingers" to describe those who are keen and have a natural talent for gardening, Thais use the term *ros-muea* (รสมือ) to highlight the special gift for cooking of an individual. The literal translation of *ros-muea* is "hand's flavor," and a person with good *ros-muea* is believed to bring any dish to the peak of perfection by achieving a point of bliss through the right balance of taste and flavor. It is probably due to this belief that cooking and seasoning are a special gift in the combination of skills. The traditional Thai cookbook seldom specifies the exact amount of each ingredient, preferring in each recipe to leave room for the cook to improvise. For example, the identity of green curry (*gaeng-keow-wan,* แกงเขียวหวาน) lies in the green color of a coconut-based curry. The primary flavor of the green curry is hot and spicy, followed by saltiness and a little sweetness (even though the direct translation of the Thai name for the dish is sweet green curry) (Bongsanid 1935; Devakula 2017; Phassakorawong 1910; UNESCO 2019).

Another fundamental reason why Thai cooks rely on their senses in judging the amount of ingredients lies in the fact that each component varies widely in its taste and aroma. For example, a teaspoon of salt yields precisely 6 grams of sodium chloride, so saltiness is consistent and predictable. However, a teaspoon of the fish sauce gives different levels of saltiness, sweetness, and aromas depending on the type of fish, methods of fermentation, etc. A teaspoon of sugar also provides consistent sweetness from 4.2 g of sucrose, while a teaspoon of palm sugar will yield sweetness, aroma, and color varieties. Other factors include seasonal, breeds, maturity, and freshness, increasing the taste, texture, and smell variation from natural food sources (details of significant ingredients are in Chapter 4).

Furthermore, most flavor compounds are sensitive to heat, and so is the ingredients' texture. Thus, the sequence of addition to reach just the right doneness is also critical to the aroma and texture of finished products. For example, the fish sauce should be added during the hard-boiling after the meat is fully cooked. The fish sauce should be added after all meat and vegetables are cooked and right before the heat source for stir-frying. Adding the fish sauce too early will result in the off-smell of the dishes (Sanitwong 1980).

With technology, the industrialization of Thai food ingredients has helped standardize the characteristics of Thai food ingredients and cooking techniques to reduce inconsistency. However, frequent testing during cooking and cooking skills are still an essential practice in cooking Thai gourmet. Therefore, the personal establishment of cooking skills is crucial and indispensable.

5.2 Moist Heat, Boiling

There are many methods of moist heat that use either water or steam as a cooking medium. The cooking temperature of the moist-heat method is naturally limited to the boiling point of water, i.e., 100°C (less if cooking at a higher altitude than sea level). Therefore, as long as the food remains moist, browning will not occur since it occurs at a temperature of at least 150°C. To prevent browning, the moisture of the food has to be maintained. The usual method to retain moisture is by covering the receptacle with a lid so that the cover prevents steam from escaping. Simmering or heating at lower than boiling point can also help to avoid the excessive evaporation of water.

The terms *gaeng* (แกง) and *tom* (ต้ม) are used to describe watery dishes. Boiling is the most common cooking method. There are many types of *gaeng* or savory soupy dishes in Thai cuisine.

The least complicated *gaeng* is a clear soup with no herbs and spices (*gaeng-jued*, แกงจืด). Other types of *gaeng* can be distinguished by the use of coconut milk, and the application of herbs and spices (Sawangsook 2021). Roles of the ingredients and technique can be conceptualized (based on details in earlier chapters) as follows.

- Coconut milk: Add creaminess and mouthfeel, intensify flavor profiles, and change the food structure to emulsion (from thin and transparent to thick and opaque liquid).
- Dry chili: The main function is to yield an orange to red color to the soup or curry. Spiciness can be toned down by removing the placenta.
- Pounding herbs and spices into a paste: Increase the release of aroma from oil glands, thicken the soup, and serve as a fundamental identity and flavor foundation for each dish. A group of ingredients required for each paste is called *kruang-gaeng*; the curry paste obtained from pounding *kruang-gaeng* is called *nam-prig-gaeng*.
- Herbs: Fresh tissue of plant parts, such as garlic, galangal, lemongrass, etc. They are high in essential oils, providing aromatic compounds to the soup.
- Spices: Dried plant parts, less aromatic but more chemesthesis, so they provide more trigeminal sensations

The matrix of Thai savory dishes is illustrated with other details in Chapter 6. The soupy dishes, cooked with the boiling technique, can be placed in boxes 5-8 (or on the right side) of the Gub-khao Grid (Fig. 6.3). There is a practically endless possibility of *gaeng* varieties, but they can be categorized as follows.

1. Clear soup: e.g., *gaeng-jued* (แกงจืด)
2. Soup with a curry paste that contains no dried chili: e.g., *gaeng-liang* (แกงเลียง)
3. Soup with a curry paste that contains dried chili: e.g., *gaeng-som* (แกงส้ม)
4. Soup with coconut milk and large visible pieces of herbs: e.g., *tom-yum* (ต้มยำ), *tom-kha* (ต้มข่า)
5. Soup with coconut milk and curry paste with no spices: e.g., *gaeng-kua* (แกงคั่ว)
6. Soup with coconut milk and curry paste with spices: e.g., red curry, green curry, massaman curry

More detail on the soup with no coconut milk (numbers 1, 2, and 3) is in topic 5.2.2, low-fat soup. More details for the curry with coconut milk (numbers 4, 5 and 6) are in topic 5.7.

5.2.1 Types of Moist Heat

The most used moist heat in Thai cooking is boiling, which is the simple act of submerging food in water and heating it until it is "done." It is the basis for making the varieties of soupy dishes that Thai people call **gaeng** (แกง). During cooking, heat transfers from the heat source through molecules of water that are in contact with each other (conduction). As they are closer to the heat source, the water molecules at the bottom of the saucepan become hotter and less dense, so they rise to the top while cooler and denser molecules on the top move down to the bottom. This liquid movement creates currents of liquid (convection) and food mass pieces that help to distribute the heat throughout the saucepan.

Boiling is used for several soupy dishes and the cooking of starchy or fibrous food materials like rice, corn, or legumes. Simmering or cooking at lower than the boiling point is used, for example, to make bone broth for vegetable soup, or to soften tough meat (by turning collagen into soft gelatin) as in spicy stews (*gaeng-hung-le*, แกงฮังเล). Simmering with little water for a shorter time is called *ruan* (รวน). This method is used in a few dishes, usually just to cook minced meat to make spicy salads such as *larb* (ลาบ).

Steaming heats food with water vapor, a method influenced by Chinese culture. The steamer consists of a perforated pan placed on top of a water-filled saucepan that generates steam. Since there is no current of hot water as with the convection of boiling, steaming is gentler and is preferred for cooking more delicate food such as fish and desserts, for instance, Thai custard and those wrapped in banana leaves. It is also used to cook starchy and fibrous food materials, especially sticky rice and tubers, pumpkin, beans, and legumes.

Blanching is another moist-heat method, but blanched food is not fully cooked, and the heating time takes only a few minutes. In Thai food, blanching is particularly crucial in preparing vegetables to be eaten as an accompanying item with chili dip. While boiling will turn the bright green chlorophyll of plants to a dull gray-green or brown, a milder method like blanching will not. When the vegetable is submerged briefly in the boiling water, oxygen, naturally stored in the plant cells (in a small structure called chloroplast), will escape. Without dissolved gas in the cells, the green color of the chlorophyll pigments becomes more visible, so blanched vegetables tend to have a brighter color. Also, with gas having escaped, the blanched vegetable decreases in volume and becomes softer. The softer and more malleable texture allows the cook to manipulate the shape of the vegetables, for example, tying long string beans into bite-sized knots for a more attractive appearance and making them easier to eat (Joachi, Schloss, and Handel 2008; Kongpan 2018). Common recipe using moist heat can be found in topics 7.2 and 7.8.

5.2.2 Low-Fat Soup

There are endless types of clear soup, main components are stock base, meat and vegetable. During boiling, the color of vegetables may change, especially those containing chlorophyll may change to brown. In addition, those vegetables with yellow to orange hues, such as pumpkin and carrot, contain fat-soluble carotenoid pigments. Hence, they may lose their pigments in the fat constituents. Another group, vegetables with a red-to-purple color, contains water-soluble anthocyanin pigments. Since anthocyanins are soluble in water, these pigments may leach out into the water. Therefore, in order to obtain as many health benefits from these beneficial phytochemicals as possible, just enough water should be used when cooking vegetables (Marcus 2013).

Meat broth, especially chicken broth, is a popular base for other soups. It can also be eaten by itself with little or no accompanying elements (vegetables, etc.). The technique of simmering at less than boiling point for a long time, similar to stewing, is called *toon* (ตุ๋น) that yields meaty, tasty soup, and very tender meat. The popular meats for this technique are chicken (*gai-toon*, ไก่ตุ๋น), duck (*pade-toon*, เป็ดตุ๋น), and pork ribs (*kra-dook-moo-toon*, กระดูกหมูตุ๋น).

The collagen component of connective tissue in meat contracts into an undesirable thick mass with severe heat, but when cooked with moist heat at a low temperature (and not allowed to reach boiling) for a long time, collagen will be denatured and transformed into water-soluble gelatin. The broth with dissolved gelatin is meaty and flavorful, and the meat will become very soft and tender. The lean muscular part of meat contains the protein actin that will be dissolved into broth also. As the meat is simmered, gelatin and actin will dissolve more into the broth along with other small peptides, free amino acids, nucleotides, minerals, and organic acids that impart the flavors' characteristics of broth and umami taste (Krasnow et al. 2012; Qi, Wang, et al. 2018). The mechanism is quite complex, especially in the changes that occur in gelatin. As gelatin is further denatured by heat, its chain structure unfolds and exposes a broader area that is ready to bind with other compounds, especially the aroma compounds that tend to share some properties (being hydrophobic or lower water solubility). The chicken broth or *gai-toon* ('ไก่ตุ๋น') is prepared with moist heat and tends to have a flavorful broth over a long time as the aroma compounds bind more closely with gelatin (Qi, Zhang, et al. 2018).

If the soup requires herbs and spices, the paste can be added to the soup base. The soup base can be water, stock, or light coconut milk, depending on the recipe. If the curry paste is used, it will be blended in the soup base before bringing it to a boil. If herbs are required in pieces, such as *tom-yum* or *tom-kha*, the lemongrass and galangal will be pounded or crushed to release the aroma before adding to the boiling soup base. The elements that require a longer cooking time will be added first, followed by those that need little heat. Finally, as mentioned earlier, certain ingredients such as fresh lime juice or fresh herbs like cilantro can be added to the serving bowl. See chapter 8 for more details recipe.

Umami taste in soup and chemesthesis of herbs and spices are discussed in Chapter 3, recipes are in Chapter 7.

5.3 Dry Heat, Grilling

The grilling technique involves the application of dry heat directly from a heat source (an open flame) to the food or direct radiant heat. Grilling is fast because the direct heat is very strong with a temperature of over 200°C. Since the traditional Thai kitchen did not include an oven, roasting, which requires a heat source at the top and the bottom of the food, was not a common cooking technique. Heat transfer by grilling is mostly radiant heat transfer from the heat source through the air onto the surface of the food. Then conduction of heat through the molecules of food will take place, and because of the high temperature of the heat source, the grilling method results in a hard, dark crust.

The grilling temperature is much higher than in boiling with lower water content, and these conditions lead to a nonenzymatic browning reaction called the Maillard reaction. Thus, not only can food turn brown, be burnt, or acquire a darker color, but the flavors and aroma of grilled food will also develop from pyrolysis, especially pyrolysis from the breaking down of fat.

Several Thai food items require grilling of some raw materials, such as chili, garlic, and shallots. For example, massaman curry requires roasting dried spices such as coriander seeds, cumin, cloves, cardamom, cinnamon, and nutmeg individually, as each is ready at a different time. The dry heating of herbs and spices results in a change of color and volatile aromatic compounds. For example, baking Thai green chili to up to 250°C for up to 30 minutes results in significant changes in color due to both the Maillard reaction and chlorophyll degradation. In addition, the burning sensation and spiciness from the capsaicin content are reduced with heat, especially at high temperatures. As for volatile compounds, aldehydes and ketones such as 3-methylbutanal, which yield a cheesy, malty odor, and hexanal, which have a grassy smell, are significantly reduced at a higher heat. At the same time, products of the Maillard reaction, such as ketones (1-octen-3-one), have a mushroom-like smell (Srisajjalertwaja et al. 2012). The roasting of black cumin seeds also results in the loss of many aroma compounds but pyrazines and furan that are proper products of the Maillard reaction increase (Kiralan 2012).

The culinary term used to describe the development of flavor and aroma from heating spices with oil is "spice bloom." The intense flavors developed with heat are then entrapped by the oil, yielding a more robust flavor (Farrimond 2017). Toasting spices is thus a valuable and straightforward technique for creating different varieties of fragrances and a variety of dishes on menus.

Most people like grilled meat with a high degree of browning for its aroma and flavor. Smoke-generated polycyclic aromatic compounds (PAHs) are absorbed in the meat. Meat with high fat content tends to have more melted fat dripping onto the coals or heat source, generating a higher concentration of PAHs. PAHs, such as benzo[a]pyrene, are carcinogenic compounds. Thus, the darker the food becomes, there is likely to be higher risk of cancer (Cheng et al. 2019; Lee et al. 2016).

Grilling in Thai cooking may involve direct heat such as the grilling of chili and shallots to prepare many chili dip dishes and the grilling of meat such as charcoal-grilled chicken, meat satay, sun-dried pork, or grilled banana. These dishes' risks contain high carcinogenic polycyclic aromatic hydrocarbons (PAHs). However, many other dishes require a barrier or a wrap before grilling, for example, *kao-larm* (ข้าวหลาม), which is the grilling of bamboo stalks containing sticky rice, sugar, coconut cream, and black beans. Banana leaves are used to wrap desserts such as sticky rice with a banana or taro filling (*kao-neow-ping*, ข้าวเหนียวปิ้ง) or meat dishes such as fish and pork and shrimp paste (*ga-pi*, กะปิ) to enhance the aroma. The bamboo or banana leaves are charred because of the intense heat. Nevertheless, these parts will be discarded, and the wrapping helps reduce carcinogens generated by grilling.

Local materials were used for cooking food in the old days. Inside bamboo stalks, sticky rice was usually served as a sweet, so sugar and coconut milk were added. In order to seal the stalk opening, the fibrous husk (mesocarp) of coconut shell or folds of dried banana leaves could be used to plug or cork the bamboo tube opening and secure the contents. Again, the fuel used to sustain flame was the dried coconut shell (mesocarp and exocarp of the coconut shell), which was practical, handy, and generated a pleasant aroma. The product could be taken on a journey and was convenient. When serving, the bamboo stalk would be broken to open, and the rice inside was covered with a thin tissue of bamboo stalk.

Thais have recognized the sweet aroma produced by burning young coconut fruit. The product is called burnt coconut fruit or *ma-prao-pao* (มะพร้าวเผา). The juice of burnt coconut fruit is sweeter. It has a characteristic smell described as a burnt aroma and burnt flavor that are desirable and potentially add value to coconut water (Jangchud, Puchakawimol, and Jangchud 2007). The burnt flavor can be concentrated by burning off coconut shells until only black ash is left. The ash will be crushed into powder and mixed with water for later use as a black colorant with a sweet burnt aroma for making a kind of dark pudding called *ka-nom-piak-poon* (ขนมเปียกปูน).

5.4 Salads

A simple mix of various ingredients creates a variety of salads, mostly spicy. The spicy salad is called *yum* (ยำ), the most well-known yum is *som-tum* (ส้มตำ). *Yum* or spicy Thai salad is a mixture of sourness, saltiness, and pungency, with a mild to no sweetness (see also topics 7.5 and 7.10). The main ingredient may be shredded green papaya, such as in the case of *som-tum* or other meat and vegetables. Because of the relatively firm texture of the green papaya, this type of salad is prepared in a clay mortar with a wooden pestle. The blunt force from pounding helps to soften green papaya as water is released from its damaged cells. More details on *som-tum* can be found in Chapter 8.

Spicy salad or *yum* is always served already mixed and tossed. There is a wide variety of dressings, and mostly the procedure starts with pounding coriander root, garlic, and chili for taste and fragrance. More garlic and chili can be added later, but with slight pounding (not to a puree consistency), so pieces are visible in the finished dish. After that, seasonings, i.e., palm sugar, sugar, lime juice, and fish sauce, are added. Other sources of sourness include tamarind paste and sour green mango. The consistency of the dressing is thin as it is fat-free but has a high intensity of taste. Some commercial yum dressings (*nam-yum*, น้ำยำ) use sucrose or palm sugar syrup to add body and thicken the dressing.

To achieve the desired texture and oral sensation, the dressing has to be prepared separately. When ready to serve, the soft and wet ingredients will be mixed first. The last ingredients that are added to the mix are those with high water content (such as fruit), and lastly, crispy ingredients such as pork rind, deep-fried dried shrimp, and toasted shallots. Tossing should be light but fast to avoid damage to the delicate vegetables and fruit ingredients.

Once tossed, the *yum* should be consumed immediately. The dressing contains a higher concentration of salt and sugar and has a lower water activity than vegetables. If the salad is left for too long, partial dehydration of the vegetables will be visible (wilted leaves) as water osmosis from vegetable cells to the surrounding dressing takes place. The osmosis is caused by the difference in osmotic pressure with water migrating from a lower osmotic pressure (vegetable cells) to a higher osmotic pressure (dressings with a higher soluble solid concentration) (Tortoe 2010). The shrinking of the vegetable cells in the dressing is due to the hypotonic of the dressing. It is the opposite of when the vegetable is soaked in water; water is hypotonic, so water osmosis into the vacuole part of the cell occurs, and, thus, the vegetables appear fresh and crisp.

Health concerns about the spicy salads in the rural areas of the north and northeastern region were due to the belief that lime juice can "cook" freshwater fish. The acid in lime juice denatures protein in raw fish, causing the color and texture to change to the color and texture of cooked fish. However, the acidity does not destroy pathogens or parasites that can be present, creating problems such as parasitic infection (Chaisiri et al. 2019).

5.5 Chili Dip

The terms chili paste and curry paste may sound similar to foreigners, but for Thai people, chili dip or *nam-prig* (น้ำพริก) is totally different from curry paste (*kruang-gaeng*, เครื่องแกง or *nam-prig-gaeng*, น้ำพริกแกง). Chili dip is prepared in a mortar and pestle, and then it can be put into a small bowl and served (see also Sections 8.6 and 8.11). Curry paste, however, is the first step in preparing a curry dish. Once a curry paste of the preferred consistency is finished in a mortar, it must be transferred to other cookware. After that, it requires the addition of liquid (mostly water and coconut milk) and heat to further the process into spicy soup dishes or *gaeng*. Chili dip is also part of the eating culture of many countries in ASEAN. It is usually prepared by pounding herbs and spices and seasoning with fish sauce and lime juice. The dish is always served with plenty of vegetables (Panyayong 2016).

The function of *nam-prig* is compatible with salad dressing as it is consumed with varieties of vegetables. The vegetables can be fresh or steamed. In addition, protein items such as hard-boiled eggs, fried sun-dried meat, fried sun-dried fish, sweetened pork, or crispy fish may be served to complement and complete the set of chili dip and vegetables. Typical ingredients used to make the chili pastes are herbs and ingredients that convey sourness, saltiness, sweetness, and spiciness. Herbs used in chili paste include garlic, shallot, kaffir lime skin, and lemongrass. Saltiness and umami may come from the fish sauce or *nam* pla, shrimp paste or *ga-pi*, fermented freshwater fish or *pla-ra* (ปลาร้า), and fermented soybean or *tua-nao* (ถั่วเน่า). Depending on what is available in season, the sourness may come from lime juice, green mango, tamarind, tomatoes, or other local small fruit and berries. The pungent, hot, and spicy sensation comes from various types of chili. Details on the pounding of ingredients, the release of aroma compounds from the plant ingredients, and the aroma of fish sauce are provided under chilies, mortar, pestle, garlic, and fish sauce topics.

Depending on the type and individual preference, the quantity of *nam-prig* consumed in a meal should be relatively low. Thus, it is the vegetables that surpass the *nam-prig* in large amounts. Vegetables are good sources of vitamins, minerals, dietary fiber, and phytochemicals (antioxidants). However, many types contain oxalate that reduces the absorption of minerals and can also precipitate into an insoluble salt. Thus, vegetables and many plant-based foods with high oxalate content may lead to mineral deficiency as well as to the formation of kidney stones (Noonan and Savage 1999).

Vegetables that tend to have high oxalate content are climbing wattle or cha-om (*Acacia pennata*, ชะอม), morning glory or *pak-boong* (*Lpomoea aquatica*, ผักบุ้ง), and bamboo shoots or *nor-mai* (*Bambusa* spp., หน่อไม้). On the positive side, the oxalate content can easily be reduced simply by boiling because the oxalate leaches out into the boiling water. The oxalate reduction is 76%, 30%, and 83%, respectively, for the *cha-om*, *pak-boong*, and bamboo shoots. These vegetables are likely to be eaten in their cooked form anyway. Rice has a very low amount of oxalate, so it is not a concern (Judprasong et al. 2006).

Another preoccupation with eating a wide variety and plenty of vegetables is the goitrogen content. Goitrogens interfere with iodine absorption, and iodine deficiency can eventually lead to goiter (enlargement of the thyroid gland). This involvement with goiter is especially true of raw cabbage, broccoli, cauliflower, and bamboo shoots (Chandra 2010). A food frequency survey in primary school students in Thailand revealed that cauliflower, peanuts, and garlic might be a concern (Mahaweerawat and Somdee 2018).

5.6 Stir-Fry

Stir-fry dishes are especially popular at lunchtime for workers during weekdays as each serving can be finished within a few minutes, compared to soup dishes, which require a longer time. This cooking technique involves using a wok and a specially shaped spatula or turning shovel (*ta-liew*, ตะหลิว).

A popular one-dish menu includes stir-fried rice such as fried rice or fried noodle, such as pad thai (ผัดไทย) and *guay-teow-pud-see-ew*

(ก๋วยเตี๋ยวผัดซีอิ๊ว). Stir-fry dishes may be served separately from a rice dish, for example, stir-fry vegetables and spicy stir-fry minced meat with basil (see also Section 8.7). The wok is needed because several turns are required per session, and the wok's depth allows the easy flipping of ingredients against its side. The ergonomics of this technique with a sloped side pan and a unique tool accommodating fast tossing distinguishes Asian style stir-frying from sautéing or pan searing. But the principles of the dry-heating of food are similar.

5.6.1 Chemical Reactions

There are many chemical reactions that occur during stir-frying resulting in a desirable brown color and the aromatic characteristics of stir-fried food. For food that contain protein and carbohydrate, when heating the food in a wok or pan until it loses moisture and is dry enough and the temperature is around 140°C, the food will start turning brown and odors will develop. Because the brown color developed from this reaction was discovered by a French researcher, Dr. Louis-Camille Maillard, the chemical reaction is called the Maillard browning reaction.

The Maillard browning is a reaction between amino acids and sugars with heat, and it is found in any food containing protein and carbohydrates. Different types of amino acids and sugars in foods yield other flavor compounds, so each food develops a unique aroma (Appell et al. 2018). Cooking with dry heat accelerates the reaction, so the Maillard browning in Thai food is also found in cooking methods such as toasting, grilling, and deep-frying (though not traditional, deep-frying has become common in the Thai kitchen). Moist-heat methods, such as boiling or steaming, maintain food temperature below the water boiling point of 100°C, so the moist food does not undergo the Maillard reaction. Food storage can also induce the Maillard browning without high heat being involved, but it takes a much longer time (Saltmarch and Labuza 1982).

On the other hand, unlike stir-frying in a wok with oil, dry-heating methods with a direct heat source such as grilling or chargrilling can expose the food to direct flame and any moisture in the food drips away quickly. This technique can easily heat food to as high as 260°C. When the temperature is higher than around 180°C, food components, mainly carbohydrates, protein, and fat, break down. This thermal degradation of food components causes the food to turn very dark or char, and an undesirable burning taste and aroma become more dominant; in other words, the food is burnt. Thus, this reaction can be called "burning" or, in chemistry terms, "pyrolysis." Thermal degradation of monosaccharides and disaccharides produces furan compounds that are then broken down further into small volatile molecules. The pyrolysis of protein, especially from meat at high temperatures of over 300°C, causes the formation of pyrolytic heterocyclic aromatic amines or heterocyclic aromatic hydrocarbons (PAHs). These compounds are carcinogenic. The carcinogens can also be found at a lower heat of around 200°C, such as frying. When fat is involved, such as grilling meat with the fat dripping onto the charcoal, more PAH, especially benzo[a]pyrene, increases (Lee et al. 2016; Lijinsky 1991; Sobral et al. 2018; Wong 2018).

5.6.2 Stir-Fried Dishes

In stir-fried vegetables, the oil is first added to the wok and heated. When minced garlic is added, a sizzling sound and garlic odor start to develop. Once the moisture in the garlic has evaporated sufficiently, brown color will develop along with the smell of cooked garlic from the Maillard reaction. However, an experienced cook will not let the garlic turn too brown because it can become bitter and smell burnt. So, just as the aroma develops and the garlic is slightly brown, ingredients that require a longer time to cook, such as meat, are added first. These are followed by ingredients that do not need so much heat, such as leafy vegetables. Lastly comes the sauce with seasonings such as fish sauce or oyster sauce. For spicy stir-fries, crushed chili and garlic are required at the start. Since fresh meat and vegetables have very high water content and stir-frying requires the rapid tossing of food in the wok to avoid uneven heat, browning from the Maillard reaction is usually slight.

Stir-fried dishes with starchy food such as stir-fried noodles (*pad-see-ew*, ผัดซีอิ๊ว; *pad thai*, ผัดไทย) or stir-fried rice (*khao-pad*, ข้าวผัด) are more susceptible to the Maillard reaction due to the lower water content of the ingredients. As a result, the contact area of the wok with food can heat to a very high level, especially when the starchy food becomes stuck to the bottom of the wok and burns. The wok used for these dishes is usually charred, and the food has the characteristic aroma of wok stir-frying (กลิ่นกะทะไหม้).

Stir-frying with flames touching the food or flash-frying at the flashpoint of cooking oil can help create a different set of aromas from steamed rice and stir-fried rice. Although it takes only 15 s of heat contact, flash-frying creates many more volatile compounds from the thermal degradation of oil than regular stir-fried rice. Pyrolysis and Maillard reactions from the protein and sugar produce some volatile compounds such as pyrazine. Still, they do not influence the fried rice aroma as the thermal degradation of fat. Separation and identification of aroma substances clearly show that steamed rice has the least number of compounds, for example, 2-acetyl-1-pyrroline with a sweet, pandan-like aroma. The aroma compounds in stir-fried rice produce smells resembling oil, stir-fried oil, and a fishy smell. Flash-frying also contributes to compounds with an odor described as, for example, scented candle, metallic, rusty, fishy/salty and sweet/stale, and the unique burnt and wok aromas. In addition, some carcinogenic PAHs are detected, though not as many (Piyachaiseth, Jirapakkul, and Chaiseri 2011). Common stir-fry recipe can be found in topic 7.7.

5.7 Curry with Coconut Milk

Curry (*gaeng-ped*, แกงเผ็ด) is a soupy savory dish, usually hot and spicy, requiring curry paste and coconut milk. Curries with distinctive colors are named, for example, red curry, green curry, and yellow curry. These curry dishes and also massaman (มัสมั่น) and *pa-nang* (พะแนง) are coconut-milk-based. There is some water-based curry, for example, *tom-yum* (ต้มยำ) and *gaeng-pa* (jungle curry, แกงป่า).

In general, curry paste is prepared from several herbs and spices, such as chili, dried chili, peppercorns, lime zest, kaffir lime leaves, kaffir lime skin, lemongrass, galangal, garlic, shallots, dried shrimp paste, coriander, cumin, cardamom, cinnamon, star anise, and cloves. The incorporation of flavor and chemesthetic agents includes the mastication of herbs and spices. The size reduction increases the efficiency of the subsequent extraction step and the physical damage of the cells activating the various enzymatic reactions of herbs and spices. Curry with coconut milk tends to have stronger trigeminal stimulation as most of the chemesthetic compounds are fat soluble, so it serves as a lipid-based extraction (Loss and Bouzari 2016). For some water-based curry or less fat curry such as *tom-yum* (spicy soup, ต้มยำ) and *tom-kha* (spicy soup with galangal, ต้มข่า), the whole or chunk of the herbs is added without grinding.

Applying mechanical force such as pounding these ingredients slowly releases the essential oils and volatile compounds. Nowadays, wide varieties of shelf-stable curry pastes are commercially available and have gained popularity, replacing the need for labor-intensive pounding with a granite mortar and pestle. Relatively clean, cutting with a sharp blade does not damage the plant cells as much as pounding with a mortar and pestle. However, many Thais still prefer to prepare fresh curry paste by pounding, claiming that a better aroma is gradually released and the consistency is more plasticity or creamy, not runny or watery as the paste obtained from motorized blades of electric blender. Their claims may be valid considering the structure of oil glands as already described in the herbs and spices section. The pounding damages a larger area of cells, squeezing the oil out rather than cutting open. Other water constituents in the vacuole and other organelles can also leach out more easily.

There are some suggestions, almost standard rules, in making curry paste. First, the dried chili has to be soaked in water till soft, and the seeds removed. Then pound large crystals of rock salt and chili, till they become very smooth, followed by coriander seed, cumin, galangal, lemongrass, and other ingredients. Shallots and garlic are the last (Bongsanid 1935). Dried chili is the first ingredient to be pounded because it is very fibrous and hard to grind. Next, salt crystals are hard and sharp so it will help to cut the fibrous chili. The step is also believed to help preventing the splashing of the chili puree. Then follow with the spices in the order of hardest to softest. Thus, shallot and garlic are the last of herbs to be added. Finally, *ga-pi* (shrimp paste) will be added last. The paste is very sticky at this point, size reduction of an ingredient is not possible after adding the *ga-pi*. The desirable consistency is the finest, no pieces of ingredient should be visible. The metaphor for the particle size is *kao-lai-mua* (เข้าลายมือ), which means the paste is so fine it can be fit into the lines on the palm of a hand.

The stiffer leaves like kaffir lime are added to the soup right after adding the curry paste. Softer herbs like sweet basil leaves will be the last added, sometimes after the heat has been turned off. The burst of flavors and odors during the crushing of the herbs and spices can be described as "taek-klin" (แตกกลิ่น). Crushing by using a mortar and pestle is preferred for a better burst of flavor. More details on the flavors of herbs and spices can be found in Chapters 4 and 5.

To cook, heat the coconut cream till the emulsion is broken and oil droplets are released. For freshly prepared coconut milk, this portion comes from the first squeeze. If processed coconut milk is used, it may be the top layer of canned coconut or simply straight out of the can with no further dilution. The cooking starts with heating the coconut cream until the emulsion breaks. Once the oil becomes visible (*ka-ti-tag-mun*, กะทิแตกมัน), the curry paste is added. The clear oil derived from emulsion breakdown is called *kee-loh* (ขี้โล้). If the coconut cream is heated in a large amount, the *kee-loh* will be achieved all at once. The too much oil layer of the curry may look unattractive. To obtain the desired amount of oil, heat just a portion of the coconut cream until the oil layer is achieved. Then add curry paste, and keep adding the remaining coconut cream little by little to prevent the curry paste from burning and maintain a desired amount of oil. The technique is called *liang-ka-ti* (เลี้ยงกะทิ). On the other hand, if less visible oil with thicker creamier soup is preferred, heat only a portion of the coconut cream at the beginning. Then add the rest to the curry when it is almost done.

Coconut milk contains high amounts of saturated medium-chain fatty acids, especially lauric acid (Bhatnagar et al. 2009). The aroma compounds are more soluble in shorter chain lipids (Guichard and Salles 2016). Hence, coconut provides good solvent properties to extract and maintain aroma compounds from herbs and spices.

After the intense aroma of curry has developed, the next ingredients are meat, followed by vegetables. At this point, the heat is reduced to simmering and the cooking continued until the desired texture has been reached. Ingredients for saltiness and sweetness can be added and adjusted to the preferred taste. Coconut milk may be heated for hours to make curry, and with amino acids and glucose, the combination of these three factors may trigger the onset of a chemical reaction called the Maillard reaction. With a greater degree of the Maillard reaction, there are concerns about the formation of a carcinogenic compound called acrylamide. However, acrylamide is formed if the curry is heated above 121°C for at least 30 minutes (Jom, Jamnong, and Lertsiri 2008). This condition is not typical in-home cooking, so acrylamide is not a concern. Common curry recipes can be found in topics 7.3, 7.4, and 7.9.

5.8 Drying

With the tropical climate suitable for varieties of fruits, Thai people have some fruits to enjoy in every season throughout the year. The fruits, for example, mango, banana, durian, pineapple, and tamarind, ripen and are ready to harvest in a massive quantity. The surplus of these fruits is preserved for year-round consumption even when the fruits are out of season.

The basic one, such as sun-drying, especially bananas, is still a regular practice. The dried banana is widely commercially available as well as exported to other countries. Important factors are the degree of ripeness at the start of drying and the final moisture content. If the drying starts with a greener banana, the texture of the finished product is firm because of high resistant starch content but not very sweet, and the aroma is very weak. On the other hand, if the banana peel is entirely yellow with visible brown spots, the texture will be too mushy as protopectin turns to soluble pectin, high sugar content, and the aroma is undesirable.

In the olden days, the end of the drying period was determined by sensory evaluation or simply guessing with experience if the bananas are dried enough. This practice was not problematic, since the reduction of moisture and the naturally high sugar content in the banana greatly reduced the chance of microorganism's growth. In other words, the dried banana has low *water activity*. Water activity is a measurement of water available for microbial growth. Therefore, the dry banana is not high risk for any foodborne illness or pathogen growth with low water activity. There is no need to add any other ingredients. Although it depends on preference, the banana can be pressed until flat or soaked in honey to enhance sweetness and health benefits.

Nowadays, other technologies such as hot air ovens are replacing the sun-drying practice. However, the processing principles and underlying mechanisms of changes remain the same. The most popular form is to dry the whole peeled banana (without any cutting or further size reduction), and the most popular variety is called **kluay-nam-wa** (กล้วยน้ำว้า). As the process continues, the banana will gradually turn brown from two chemical reactions: enzymatic browning and Maillard reactions.

Enzymatic browning reaction starts when the peeled fruit is exposed to oxygen, such as when the fruit is cut or peeled. The enzyme polyphenol oxidase (PPO) will oxidize the substrate, i.e., polyphenolic compounds, and produce brown pigment of melanin. This browning or darkening of fruit color is the same as what occurs in cutting apples or potatoes. The other browning reaction, called the Maillard reaction, does not involve an enzymatic reaction. Factors contributing to the browning from the Maillard reaction are sugar and protein content, which intensify with heat and time. In the case of dried bananas, the majority of brown pigment is hydroxymethyl-furfural (HMF), indicating that the Maillard reaction is the primary browning reaction. Thus, it is typical for a dried banana to have a brown color. However, there are several possible techniques to reduce the intensity of brown color, such as dipping in 500 ppm citric acid or 1.0% ascorbic acid for 15 minutes before drying (Auan-on 2003; Sudprasert 2003).

To extend the shelf life of the final product, its water activity has to be controlled to less than 0.85 in order to prohibit or slow down the growth of microorganisms. With a high concentration of sugar, this standard is also adopted by the Thai Industrial Standards Institute (TISI), Ministry of Industry to apply to all scales of the dried banana production (Ministry of Industry, Thai Industrial Standards Institute 2015).

5.9 Conclusion

For Thai people, soup is merely a description of any dish with a lot of water and requires a bowl for serving. The name, which begins with *tom* or *gaeng* (ต้ม, แกง), clearly defines this character and its cooking technique of boiling. On the other hand, the dishes with more solid mass are either cooked with heat, such as those stir-fried dishes, or without heat, such as those salad-style *yum* (ยำ) dishes. The use of herb and spice mixtures gives an endless possibility to create recipes. While meat and vegetable ingredients can vary widely and are substitutable, the cooking methods and the blend of herbs and spices are not at the same degree of flexibility.

Chapter 4 has pointed out that people have a definite mindset of how a dish should taste, smell, feel, and look. Chapter 5 laid out how diverse the sensory and physicochemical properties of each ingredient can be. Then, the current chapter explains how complicated and challenging it is to transform so many raw materials within a short period of cooking time into a dish that meets everyone's standards. While the expected palatability of a dish is very judgmental, the ingredients' availability and cooking techniques are not as systematic. Thus, a cook will have to be very resourceful and facilitative. In the view of Thai people, success in any kitchen seems to rely intensely on an individual, not any golden recipe or culinary code. Consequently, good cooks were renowned for their gifted talent (*ros-muea-dee*, รสมือดี) rather than for their knowledge or experience.

References

Appell, Michael, Hurst, W. Jeffrey, Finley, John W., de Man, John M. 2018. "Amino Acids and Proteins". In *Principles of Food Chemistry*, Springer, Cham (2018), pp. 117–164.

Auan-on, Tanut. 2003. *Process and Quality Development of Dried Banana*. Kasetsart University. http://fic.nfi.or.th/knowlegdedatabankResearch-detail.php?id=867.

Bhatnagar, Ajit Singh, P. K. Prasanth Kumar, J. Hemavathy, and A. G. Gopala Krishna. 2009. "Fatty acid composition, oxidative stability, and radical scavenging activity of vegetable oil blends with coconut oil." *Journal of the American Oil Chemists' Society* 86 (10):991–999.

Bongsanid, Yaovabha. 1935. *Tumrub Sai Yaowapa (ตำรับสายเยาวภา)*. Bangkok, Thailand: Saipunya Sakakom.

Chaisiri, Kittipong, Chloé Jollivet, Pauline Della Rossa, Surapol Sanguankiat, D Wattanakulpanich, Claire Lajaunie, Aurélie Binot, M Tanita, S Rattanapikul, and D Sutdan. 2019. "Parasitic infections in relation to practices and knowledge in a rural village in Northern Thailand with emphasis on fish-borne trematode infection." *Epidemiology & Infection* 147.

Chandra, Amar K. 2010. "Goitrogen in food: Cyanogenic and flavonoids containing plant foods in the development of goiter." In *Bioactive Foods in Promoting Health*, edited by Ronald Ross Watson and Victor R. Preedy 691–716. Elsevier.

Cheng, Jiali, Xianhui Zhang, Yan Ma, Jia Zhao, and Zhenwu Tang. 2019. "Concentrations and distributions of polycyclic aromatic hydrocarbon in vegetables and animal-based foods before and after grilling: Implication for human exposure." *Science of The Total Environment* 690:965–972. doi: https://doi.org/10.1016/j.scitotenv.2019.07.074

Devakula, Kwantip. 2017. *Eating in 2017*, edited by Kiraya Leksomboon. Bangkok, Thailand: Thailand Creative & Design Center.

Farrimond, Stuart. 2017. "The science of herbs, spices, oils, and flavoring." In *The Science of Cooking*, edited by Stuart Farrimond, 186–187. London, Great Britain: A Penguin Random House Company.

Guichard, Elisabeth, and Christian Salles. 2016. "Retention and release of taste and aroma compounds from the food matrix during mastication and ingestion." In *Flavor*, edited by Patrick Etiévant, Elisabeth Guichard, Christian Salles and Andrée Voilley, 3–22. Elsevier.

Jangchud, Kamolwan, Pimolpan Puchakawimol, and Anuvat Jangchud. 2007. "Quality changes of burnt aromatic coconut during 28-day storage in different packages." *LWT-Food Science and Technology* 40 (7):1232–1239.

Joachi, Daivd, Andrew Schloss, and Philip A. Handel. 2008. *The Science of Good Food; the Ultimate reference on how cooking works*. Canada: Robert Rose Inc.

Jom, Kriskamol Na, Pimon Jamnong, and Sittiwat Lertsiri. 2008. "Investigation of acrylamide in curries made from coconut milk." *Food and Chemical Toxicology* 46 (1):119–124.

Judprasong, Kunchit, Somsri Charoenkiatkul, Pongtorn Sungpuag, Kriengkrai Vasanachitt, and Yupaporn Nakjamanong. 2006. "Total and soluble oxalate contents in Thai vegetables, cereal grains and legume seeds and their changes after cooking." *Journal of Food Composition and Analysis* 19 (4):340–347.

Kiralan, Mustafa. 2012. "Volatile compounds of black cumin seeds (*Nigella sativa* L.) from microwave-heating and conventional roasting." *Journal of Food Science* 77 (4):C481–C484.

Kongpan, Srisamorn. 2018. *Intangible Cultural Heritage Foods of Thailand* (อาหารขึ้นทะเบียน มรดกทางภูมิปัญญาของชาติ) Bangkok, Thailand: S.S.S.S. (บริษัท ส.ส.ส.ส. จำกัด).

Krasnow, Mark, Tucker Bunch, Charles Shoemaker, and Christopher R Loss. 2012. "Effects of cooking temperatures on the physicochemical properties and consumer acceptance of chicken stock." *Journal of Food Science* 77 (1):S19–S23.

Lee, Joon-Goo, Su-Yeon Kim, Jung-Sik Moon, Sheen-Hee Kim, Dong-Hyun Kang, and Hae-Jung Yoon. 2016. "Effects of grilling procedures on levels of polycyclic aromatic hydrocarbons in grilled meats." *Food Chemistry* 199:632–638.

Lijinsky, William. 1991. "The formation and occurrence of polynuclear aromatic hydrocarbons associated with food." *Mutation Research/Genetic Toxicology* 259 (3):251–261. doi: https://doi.org/10.1016/0165-1218(91)90121-2

Loss, Christopher R, and Ali Bouzari. 2016. "On food and chemesthesis—Food science and culinary perspectives." In: *Chemesthesis: Chemical Touch in Food and Eating*, edited by Shane T. McDonald, David A. Bolliet, and John E. Hayes, 250. doi: https://doi.org/10.1002/9781118951620.ch13.

Mahaweerawat, Udomsak, and Thidarat Somdee. 2018. "Iodine fortification of dessert in iodine deficiency prevention program for primary school children, Maha Sarakham Province, Thailand." *Southeast Asian Journal of Tropical Medicine and Public Health* 49 (3):502–508.

Marcus, Jacqueline B. 2013. *Culinary Nutrition: The Science and Practice of Healthy Cooking.* Academic Press.

Mauer, Lisa J.; San Martin-Gonzalez, M. Fernanda. 2017. "The principles of food science." In *Culinology; the Intersection of Culinary Art and Food Science,* edited by J. Jeffrey Cousminer, 19–45. Canada: John Wiley & Sons.

Ministry of Industry, Thai Industrial Standards Institute. 2015. MPH.112/2558. edited by Ministry of Inudstry. Thailand.

Noonan, S. C., and G. P. Savage. 1999. "Oxalate content of foods and its effect on humans." *Asia Pacific Journal of Clinical Nutrition* 8 (1):64–74.

Panyayong, Chunkamol. 2016. "Study of ASEAN food culture for ASEAN soci-cultural community." *Institute of Culture and Arts Journal* 18 (1):43–52.

Phassakorawong, Thanphuying Plian. 1910. "Tamra maekhrua hua pa." In, ed Plain Phassakorawong. Bangkok: Tonchabap. https://vajirayana.org.

Piyachaiseth, Tunyatorn, Wannee Jirapakkul, and Siree Chaiseri. 2011. "Aroma compounds of flash-fried rice." *Kasetsart Journal: Natural Science* 45:717–729.

Qi, Jun, Hu-hu Wang, Wen-wen Zhang, Shao-lin Deng, Guang-hong Zhou, and Xing-lian Xu. 2018. "Identification and characterization of the proteins in broth of stewed traditional Chinese yellow-feathered chickens." *Poultry Science* 97 (5):1852–1860.

Qi, Jun, Wen-wen Zhang, Xian-chao Feng, Jia-hang Yu, Min-yi Han, Shao-lin Deng, Guang-hong Zhou, Hu-hu Wang, and Xing-lian Xu. 2018. "Thermal degradation of gelatin enhances its ability to bind aroma compounds: Investigation of underlying mechanisms." *Food Hydrocolloids* 83:497–510.

Saltmarch, Miriam, and Theodore Labuza. 1982. "Nonenzymatic browning via the maillard reaction in foods." *Diabetes* 31 (1982):29–36. doi: 10.2337/diab.31.3.S29

Sanitwong, Mom Ractchawongse Tuang 1980. "The important things to know (สิ่งสำคัญที่ควรทราบ)." In *Tumrub Sai Yaowapa (ตำรับสายเยาวภา),* edited by Yaovabha Bongsanid, 34–35. Bangkok, Thailand: Saipunya Samakom.

Seubsman, Sam-ang, P. Suttinan, Jane Dixon, and Cathy Banwell. 2009. "20 – Thai meals." In *Meals in Science and Practice,* edited by Herbert L. Meiselman, 413–451. Woodhead Publishing.

Sobral, M Madalena C, Sara C Cunha, Miguel A Faria, and Isabel MPLVO Ferreira. 2018. "Domestic cooking of muscle foods: Impact on composition of nutrients and contaminants." *Comprehensive Reviews in Food Science and Food Safety* 17 (2):309–333. doi: 10.1111/1541-4337.12327.

Srisajjalertwaja, Siriwan, Arunee Apichartsrangkoon, Pittaya Chaikham, Yasinee Chakrabandhu, Pattavara Pathomrungsiyounggul, Noppol

Leksawasdi, Wissanee Supraditareporn, and Sathira Hirun. 2012. "Color, capsaicin and volatile components of baked Thai green chili (*Capsicum annuum* Linn. var. Jak Ka Pat)." *Journal of Agricultural Science* 4 (12):75.

Sudprasert, Sirirat. 2003. "Control of Browning in Dried Banana *Musa sapeintum* L. Product (การ ควบคุม การ เกิด สี น้ำตาล ใน ผลิตภัณฑ์ กล้วย *Musa sapientum* L. ตาก)." Food Technology, จุฬาลงกรณ์ มหาวิทยาลัย.

Tortoe, Charles. 2010. "A review of osmodehydration for food industry." *African Journal of Food Science* 4 (6):303–324.

UNESCO. 2019. "Intangible Cultural Heritage." https://ich.unesco.org/en.

Wong, Dominic WS. 2018. "Natural toxicants." In *Mechanism and Theory in Food Chemistry*, edited by Dominic W. S. Wong, *Second Edition*, 327–359. Springer.

PART III

THE CULINARY INTEGRATION OF ART AND SCIENCE

6

EATING PLEASURE OF THAI MEAL

VALEERATANA K. SINSAWASDI, HOLGER Y. TOSCHKA, AND NITHIYA RATTANAPANONE

Contents

6.1 Introduction

The first Thai cookbook written by a foreigner was Siamese Cookery. It was written by an American author, Marie Wilson. (Wilson 1965). Wilson lived in Thailand in the 1950s and found the greatest pleasure in the hot and spicy flavors of the cuisine, new tastes, smells, sights, and sounds defined the encounters, the spiciness or hotness of some dishes awoke new parts of the palate for the culinary adventurers. Another cookbook, also by an American, was written by Brennan (1981). In Brennan's view, *Thai food awoke all the human senses, unusual aroma, fragrant jasmine to pungent shrimp paste, a variety of textures* (Padoongpatt 2017). In addition, as described in Chapter 1 and more foreigners' comments as mentioned in Appendix A, Thai foods can bring joy to just about everyone, from the first try of *tom-yum-koong*, or repeated consumption, to the frequent or habitual dish of the local Thais.

The intricacies of food and eating culture in the form of meal settings are a unique tradition of Thailand. A Thai meal cannot be standardized or quantified sense by sense, but all elements are holistically elaborated into a work of art. The assortment of dishes has a fundamental pillar on balance yet customizable sensory modalities: appearance, taste, smell, sound, texture, and even mild irritation or pain. From umami and basic tastes to the aroma and irritants of herbs and spices, there are several fascinating layers of the unique Thai cuisine's deliciousness. The rice is like a white space of artwork that can either accentuate or dilute down any attributes of the meal. In this chapter, the integration of sensory cues within each dish and the eating pattern is analyzed with scientific theories to discuss the reason behind the affection of this national culinary tradition.

6.2 The Thai Way of Eating

Traditionally, a Thai meal consists of rice, fish (or other meat dishes), chili dip (*nam-prig*, น้ำพริก), and a lot of vegetables and herbs. This combination is usually referred to as "*khao and gub-khao*" (ข้าวและกับข้าว), meaning "rice and savory add-ons." There is a wide variety of gub-khao (กับข้าว). The most common cooking techniques of gub-khao are as follows: simmering curry, stir-frying (with or without chili), spicy mixing (yum or spicy salad), frying/broiling/grilling, crushing, and grinding (the hot and spicy dipping sauce) with a mortar and pestle, and condiments such as pickles and salted fish. There are 3-5 dishes in a meal served all at once. The pattern may be comparable to a 5-course Western dinner but lay all items out at once as if they are a single course. This unique ensemble is called *sum-rub*. (Nitivorakarn 2014; Seubsman et al. 2009).

In the past, Thai people ate on the floor in a circle with shared dishes (and common spoons) in the middle. They used their fingers to manipulate food on the plate and to deliver it to the mouth. Flatware or utensils for individual users were not introduced to the dining table until around the 1870s. However, as rice is the staple food, their grains are too small to be handled with a fork and knife. Spoon is best to hold rice together with other semi-solid and liquid food (Plainoi 2015). The knife had, thus, been omitted as cutlery on the dining table and it was used as a kitchen tool only. In Thai dishes, meat and other tough items are almost always prepared to a bite-sized before serving, eliminating the need for a knife while eating. Thus, the role of the fork is reduced to only push the food onto the spoon. With these different cutlery functions, the unique pattern has been developed into national tradition, i.e., a spoon is held by the right hand and a fork on the left. The size of the fork and spoon has been adjusted over time for the right balance and feel during the meal. The spoons and forks are of the same size (lengthwise), very similar weight, and are sold in pairs. Thai people use flat plates for rice, not chopsticks and bowls.

Thai *sum-rub* of *gub-khao* varieties is meant to be shared. On the dining table, there is a plate of rice in front of each diner, the *gub-khao* dishes in the middle, and a communal spoon accompanying each *gub-khao*. When ordering from a menu in a restaurant, it is not necessary to order the same number of dishes as the number of people as each one is to be shared.

6.3 Profound Impact of Chili Pepper

The touch of capsaicin on the tongue gives a unique sensation of what may be described as hot, fiery, burning, stinging, pungent, spicy, and piquant. This sensation is not really a taste as the capsaicin doesn't bind with the taste receptor on the tongue but with sensory neurons that sense heat and pain signals to the brain (Smutzer et al. 2018). It is not difficult to tell if someone is eating spicy food, with a flushing face, runny nose, teary eyes, sweating, breathing through their mouths fast, and sometimes fanning their hands in front of their open mouths in an attempt to cool down the burning sensation. These body reactions are the responses to flush the irritant away from the sensory neurons in the oral and nasal cavities. A large dose of hot spicy food may also cause nausea and vomiting (Lawless, Rozin, and Shenker 1985; Srinivasan 2016). These body reactions give an impression of a person in distress and are suffering from eating spicy food, yet chili is an indispensable ingredient in many Thai dishes.

6.3.1 Chili in Thai Food

The first two Thai cookbooks dated 1890 and 1908 discuss curry paste and chili in a significant proportion of the recipes and are usually signified by the words *prig* (chili) and *ped* (spicy hot), including in the name of the dishes. This fact reflected how much Thai people have loved the piquancy of the chili (Namwong and Pewporchai 2019; Pewporchai 2017). Therefore, it's not surprising to come across comments such as "Thai food without chili is not a Thai food" or a claim that chili is the most vital Thai cooking ingredient (Sukphisit 2019). For local Thais, this hot burning sensation is called *ped* (เผ็ด). The phrase *mai ped mai aroy* or "if it's not burning hot, it's not delicious" makes a fun small talk topic on dining table.

For many Thai people, the *mai-ped* (not spicy hot) does not mean the absence of chili, but just signaling a reduction of the chili peppers in the dish. Those customers asking for *som-tum* (spicy papaya salad) with "no chili" may meet with a confusing look from the cook or waiter. One common explanation from the kitchen is that if chili is completely excluded, the flavor is impaired beyond the point of being a *som-tum*.

At a glance, it seems irrational to continue eating spicy food despite all the signals of body suffering that seem to linger even after fininshing the meal. However, chili has become an essential food ingredient to those local Thais who consume it regularly. Those foreigners who try the spicy dish for the first time easily develop the liking and craving for spicy food afterward (see also Chapter 3, under the topic of pungency).

In general, local Thais are exposed to spicy dishes early in life. The flavor of regional cuisine may already influence how we like the food (Schaal, Marlier, and Soussignan 2000) and gradually increase the degree of spiciness with time. As consumption increases, the sensitivity to the burning sensation decreases. In other words, those who consume chili regularly can lead to a chronic desensitization, which causes them to not feel as much burning sensation. The reduction in the burning intensity ratings of spicy dishes can be observed in just after two weeks of repeated exposure, i.e., more chili is required to obtain the same level of spiciness of a dish. Those who like spicy food tend to acquire more spiciness (Carstens et al. 2002; Dessirier et al. 2000; Ludy and Mattes 2012; Stevenson and Prescott 1994).

For Thai people, chili's hotness can go well with just about any food item and eaten on any occasion to elevate their flavor profile. Chili is essential to many condiments of any profile, salty, sour, or sweet. Examples are as follows.

- Fresh chili slices in fish sauce (*prig-nam-pla*, พริกน้ำปลา), sometimes with sliced garlic or shallot. It is the most common condiment, almost always a must when serving rice.
- Fresh chili, sliced or pounded, pickling in vinegar is known as sour chili (*prig-nam-som*, พริกน้ำส้ม). This tangy and spicy condiment is essential for noodle dishes, whether soup or stir-fried. The chili vinegar is commonly placed in a four-compartment container customarily with fish sauce, sugar, and dried chili.
- Any flavor of instant noodle, even a plain minced pork flavor with no spices for kids, always contains a small sachet of dried chili in the package.
- Dried chili mixed with salt and sugar, known as *chili and salt* (*prig-gub-kluar*, พริกกับเกลือ), is a condiment item paired with any fresh or pickled sour fruits.
- A thick gluey dipping sauce of palm sugar boiled with fish sauce, shallot, dried shrimps, and fresh chili slices is called *sweet fish sauce* (*nam-pla-waan*, น้ำปลาหวาน). It is specially paired with sour green mango.
- A relatively new chili mixture popular with various fruits products (fresh, pickled, and dehydrated) is the chili and plum powder (*pong-buay*, ผงบ๊วย).
- Clear syrup-based dipping sauce with pieces of minced chili known as sweet Thai-style dipping sauce (*nam-jim-gai*, น้ำจิ้มไก่). Other spicy dipping sauces with more complicated ingredients are, for example, Sriracha sauce and seafood dipping sauce.

6.3.2 Spiciness Imparts Deliciousness, What Else?

For foreigners who were irritated by the capsaicin but turned to like it eventually, the reasons may include the masochistic/thrill-seeking hypotheses. Like riding a roller coaster in a theme park, or watching sad movies, the joy comes from endogenous opioids released when exposed to what can be perceived as a danger but cannot really cause any actual harm (benign masochism). In addition, many studies pointed out that certain personality traits, such as sensation seeking, possibly cause some people to enjoy the burning sensation of chili (Byrnes and Hayes 2013, 2015; Cooke and Fildes 2011; Dalton and Byrnes 2016; Lawless, Rozin, and Shenker 1985; Rozin and Schiller 1980; Schaal and Durand 2012; Spence 2018; Spinelli et al. 2018; Terasaki and Imada 1988). Interestingly, eight out of ten among the top ten Thai dishes that foreigners like to order from Thai restaurants overseas, such as *tom-yum-koong* and green curry (see Chapter 1), are spicy dishes that contain chili (Chavasit et al. 2003a).

In addition to the endogenous opioids released from the benign threat, the sensory neurons that bind with the capsaicin also send the pain signal to the brain. To reduce the pain, the brain (pituitary gland) secretes endorphins, which are endogenous opioids, and increases dopamine (Dishman and O'Connor 2009; Sprouse-Blum et al. 2010). These chemicals bring pleasantness and joy; this can also explain why people like spicy food despite its burning and stinging sensation (Choy, El Fassi, and Treur 2021).

The capsaicin in chili can cause desensitization, which can inhibit flavor perception for several minutes, after which the food being eaten tends to taste bland (Hayes 2016). This change in perceivable burning sensation may cause by the adaptation of the sensory neurons. Since the perception of heat is reduced, the secretion of endorphin and dopamine is also reduced. Thus, many people tend to increase the level of spiciness in the food or eat more spicy food so that they can obtain the same level of pleasure (Touska et al. 2011). However, spicy food does not cause overeating. Capsaicin in the diet is likely to increase satiety, thereby reducing overall energy and fat intake (Westerterp-Plantenga, Smeets, and Lejeune 2005). In a large-scale study from China, it is observed that those who regularly consumed spicy food had better blood lipid profile indicating a lower risk of cardiovascular diseases (Xue et al. 2017).

With several pungent spices on the menu, there are other possible sensations pertaining to the trigeminal nerve stimulation, such as burning and tingling reaction from the cinnamaldehyde in cinnamon; burning, tingling, and numbing from the cuminaldehyde in cumin; long-lasting numbing from the eugenol in cloves and holy basil, and burning from the piperine in black pepper and ginger oleoresin in ginger (Cliff and Heymann 1992). After the meal, the lingering feeling in the mouth largely depends on the burning sensation of herbs and spices, especially chili, which can be a nuisance and unpleasant. The magnitude of the burning sensation can temporarily be reduced when other food is in the mouth but comes back again when the mouth is empty. Suggested approaches to reducing this sensation are as follows (Hutchinson, Trantow, and Vickers 1990; Lawless and Heymann 2010; Lee and Kim 2013; Samant et al. 2016):

- Eating rice: Starchy food like rice and corn, colder food and cold water, and anything with a sweet taste work well to mitigate chili burn.
- Taking a sour sip: Sourness increases saliva flow in the mouth, so it can help to reduce chili burn.
- Drinking milk: Milk has been shown to be best compared with Oolong tea, tomato juice, and sparkling and spring water for cleaning the palate after spicy chicken. The strategy works well both initially and over the course of 2 minutes.
- Eating a dessert: A combination of fat and sugar is also effective in an experiment where milk with a higher fat content works well, especially with some added sweetness.

Since capsaicin is fat-soluble, foods with fat and oil potentially lessen the burn. Though a Thai meal *sum-rub* may offer many spicy dishes, curry, chili paste, spicy salad, etc., Thai people have a habit of neutralizing their palette with vegetables. Also, the Thai culinary wisdom of pairing food with specific accompanying dishes makes the dining experience very dynamic. After the savory course, various Thai desserts, mostly made of coconut milk and sugars, can further help to wash away pungent substances and help the taste and smell receptors to recover fully. Thai people like to end the course of a meal with fruit. The phrase used to describe this practice is "eating fruit to refresh the palate," which likely refers to the refreshing sensation as the burning, numbing, and tingling sensation is thoroughly washed away.

6.4 Case Study: Meal Replacement Product for Thai Consumers

In early 2000s, the author, Valeeratana, was a Research and Development Manager specifically responsible for formulating the meal replacement products for weight loss purposes. As part of a preliminary evaluation program, various formats of the meal replacement products were imported from the Unilever USA for in-house volunteers (Unilever Thailand employees). To lose weight and maintain good health, each volunteer had to replace their regular meals with meal replacement products twice daily.

There were dairy-based ready-to-drink products in metal cans (shakes) and ready-to-shake powder products designed to be mixed with low-fat milk (shake mixes). Besides the sweet product profile, there was savory cream soup prepared simply by adding hot water. Each format had varieties of flavors. However, all products had the same energy range of 220–250 kcal per serving, good quality protein (PDCAAS of over 0.90), and one-third %RDI of all micronutrients. While the results were satisfying initially as volunteers were happy with their weight loss progress, the product development team started to receive repeated complaints from participants that they craved spicy food. Even though the weight loss plan allowed one regular meal a day of the participants' choice, the one meal allowed seems to be unfulfilling.

At the time, this meal replacement product had been on the large-scale markets for over 20 years. The products were popular in many Western countries, and the company had never been aware of this consumer comment elsewhere besides Thailand. This craving could not be satisfied with the existing western-style savory product, such as creamy chicken soup. The Marketing and R&D teams took this request as crucial Thai consumer insight and formulated a spicy meal replacement product range. Product acceptance was of utmost importance to prevent early withdrawal of participants from the weight loss program.

After several market and consumer research studies and several sensory tests, four Thai flavor instant noodle meal replacement prototypes were selected for commercialization. The flavors include *woonsen tom-yum* (วุ้นเส้นต้มยำ), *woonsen suki* (วุ้นเส้นสุกี้), *pasta kee-mao* (พาสต้าขี้เมา), and *pasta rad-na* (พาสต้าราดหน้า) (Fig. 6 a-b-c-d). During the product development process, the team had learned more consumer insights. For example, most Thai people liked very spicy food (several rounds of increases in spiciness were required during prototype development). Out of these four flavors, *pasta rad-na* was the only one that contained no chili, and it was the least popular (the least requested for during the weight loss program) among the four. Not just the flavor, the texture was critical and was considered another consumer insight. Hence, the team developed two soup varieties (*tom-tum* and *suki*) and the other two with non-soup consistency (*kee-mao* and *rad-na*).

When spiciness from chili was the sensory dominance, it was easier to develop the product to the satisfaction of prospective consumers. Perhaps, it is also because the chili spiciness really enhances desirable flavors or chili's desensitization property reduces the perceived intensity of fortified vitamins and minerals. Another challenge was in the development of the spicy stir-fry flavor of holy basil (*pad-ka-prao*, ผัดกะเพรา). It was a challenge to capture the genuine holy basil aroma in a shelf-stable product. Hence, the *pad-kee-mao* (ผัดขี้เมา), with sensory dominance of spiciness from chili and pungency from dried spice like black pepper, gained more preference during development since there is no noticeable off-flavor. Typically, the essential oils from dried spices such as black pepper are more stable than those from fresh herbs.

The most challenging flavor was the pasta *rad-na* (พาสต้าราดหน้า). With no spiciness to draw the sensory attention to, or pungent herbal scent to mask off-flavors (including from fortified vitamins and minerals), the focus was on the characteristic smell, i.e., the aroma of burnt food at the bottom of a wok (กลิ่นกะทะไหม้, see Section 6.6.2). The flavor house spent quite a long time studying the volatile compounds of this stir-fried dish using gas-chromatography-olfactometry (GC-O). Finally, a success, the artificial flavor, was explicitly developed to mimic the burnt-wok aroma in *rad-na*. Taste result showed the *rad-na* got significantly higher preference than the pork stew flavor (moo-toon, หมูตุ๋น), albeit *rad-na* was the least popular compared to the other three flavors, which were all spicy.

After an in-house evaluation, large-scale consumer research was conducted by an independent research agency. The results from both

Figure 6.1   Asian-Range Slim-Fast meal replacement product prototype. (a) Woonsen *tom-yum*. (b) *Woonsen suki*. (c) Pasta pad *kee-mao*.

Figure 6.1   Asian-Range Slim-Fast meal replacement product prototype. (d) Pasta *rad-na*.

consumer research studies share similar findings. Finally, these meal replacement products were used in a weight management study in over-weight and obese women in collaboration with Mahidol University Faculty of Medicine Ramathibodi Hospital (Lepananon 2005). All meal replacement products were formulated and produced in Thailand, including ready-to-drink shakes (strawberry, coffee, and choco-late), shake mixes (vanilla and chocolate), and instant noodles (*suki, tom-yum, kee-mao,* and *rad-na*). Thirty-five subjects participated in this meal replacement plan for 36 weeks (from 85 participants total for all tested weight loss plans combined). Consumption behaviors identified in this study confirmed that Thai people crave spicy food, especially for lunch and dinner. However, the spiciness was not expected for break-fast. As a result, the savory and spicy instant noodles were consumed more as a second meal of the day. At the end of the program, partici-pants lost an average of 6.9 kg (8.6% of body weight).

The results of the consumers' attitudes study and weight loss strategy evaluation linked spiciness to sensory-specific satiation at lunch and dinner. An offering of western-style savory products, such as a cream soup, did not help elevate the satiation (the satisfaction of completing a meal or feeling full). Interestingly, the hot burning sensation from chili seemed to trigger the satiation. However, the spiciness did not increase appetite or promote more caloric consumption either. In the meal-replacement weight loss plan, participants actually lost weight and maintained the weight during a 36-week study period. After the registration of 17 products with Thailand FDA, large-scale production runs, TV commercial shoots, and launch date set, the company abruptly decided to terminate the project in all Asian markets. The decision came as the global team predicted the declining trend worldwide on this weight loss strategy. Still, this project's findings highlighted the significance of spicy food to Thai people.

6.5 *Gub-khao*: An Innovation Turns Heritage

Understanding how to create a menu for each meal and how to serve it is essential to gain a positive experience of eating a Thai meal. Singsomboon (2015) observed that, unlike Japanese cuisine, where foreigners capture the eating tradition quite well, Thai restaurant customers outside Thailand or foreign visitors to Thailand do not understand the fundamentals of composing a meal out of the choices available on the menu. These common misunderstandings potentially decrease the joy of a meal. For example, eating a one-plate dish like pad thai with rice turns the meal into a monotonous and unflavorful one. It is better to order an assortment of savory à la carte (*gub-khao*, กับข้าว) from a menu in fine dining, which will elevate the total experience. On the other hand, an unbalance of flavors and sensations between dishes can ruin the whole meal. For example, if all plates are spicy, there will be an overpowering hot burning sensation on the palette without a break for refreshing. This makes it impossible for anyone to enjoy the meal. Furthermore, some foreigners do not understand the concept of sharing assorted dishes and think of each one as an individual dish. Therefore, a better understanding of Thai eating culture will enable more foreigners to enjoy the full potential of the Thai meal experience.

6.5.1 *Multisensory Properties of Savory Gub-khao (Thai Dishes)*

Thai meals provide a sensory experience that stimulates all sensory receptors and evokes all human senses. There is a colorful and thoughtful presentation of the different combinations of flaovrs and texture of food. At the very high end, there is what has become known as the Royal Thai cuisine, and here the esthetic aspect of food is even more prominent. The skillful carving and preparing each ingredient are time-consuming but mesmerizing and stimulate the sense of sight. As the saying goes, "eye appeal is half the meal." Presentation of food, especially the more formal meal, is crucial. Vegetable and fruit carving, to serve as a decoration, a garnish, or as food pieces to be eaten, is another mastery passed down for generations. An article in The Spruce Eats (https://www.thespruceeats.com) a foodie website that boasted 243 million readers annually and more than 35,000 international food photos posted, admires the Thai dining table decoration as *Thai food presentation is among the most exquisite in the world* (Schmidt 2021).

Thailand has an endless list of raw materials to cook into a dish. But the Thai dining experience is more than just what the eyes and nose perceive. Incorporating spices and herbs into recipes adds dimensions, depth, and enhancing or masking certain flavors and senses, boosts deliciousness, and most notably contributes to the characterization of Thai food. The extra flavors, taste, aroma, and chemesthetic agents stimulate the sense of touch and trigeminal resulting in uniquely delectable food. Other countries in Southeast Asia, such as Malaysia, Vietnam, and China, also have their curry dishes. However, varieties of Thai curry are uniquely prepared from a moist paste of herbs and spices, while others are mostly made with a mixture of dry spices. For example, typical massaman curry contains six herbs and ten spices in its curry paste. Green curry, another famous Thai curry, requires nine herbs and three spices.

Generally, herbs are referred to plant leaves, while spices are referred to dried seeds, berries, roots, or bark of plants. They are not only providing complex, volatile compounds but are also good sources of chemesthetic agents. The irritation or pain sensations from the chemesthetic effect include heating, burning, cooling, tingling, numbing, pungency, mouth-puckering, and astringency. These chemesthetic agents are polyphenolic compounds of wide varieties. Each can evoke trigeminal system by a single active compound, or interact with proteins, resulting in many sensations, especially the astringency (García-Estévez, Ramos-Pineda, and Escribano-Bailón 2018).

The combinations of these sensory experiences are thought to create a craveable sensory experience. Other cuisines require spice mixture, i.e., Southern United States (Louisiana), Morocco, Argentina, Egypt, China, India, West and South Africa, Italy, France, Mexico, Japan, and the Middle East. However, Thai curry paste is the only one that contains fresh lemongrass and kaffir lime leaf, shallot, and galangal (Danhi and Slatkin 2009; Haley and McDonald 2016; Loss and Bouzari 2016). In addition to plant-based ingredients, Thai curry paste also contains fermented animal products like shrimp paste (*ga-pi*, กะปิ). The flavor compounds in fresh herbs add the invigorating or refreshing note to the finished product, while the seasoning like shrimp paste and fish sauce give the umami and kokumi taste that can linger in the mouth (see Section 4.2.3).

The unique combination of ingredients and an ingenious preparation and cooking method gives rise to the identity of each Thai dish. The mixture of herbs and spices, in particular, is responsible for the identification and recognition of the finished product. This inability to single out any specific source of flavor reflects the fact that human senses cannot effectively distinguish the taste and odor of each ingredient out of a mixture (Jinks and Laing 1999; Laing et al. 2002; Marshall et al. 2005). Considering the possibility of flavor pairing (Ahn et al. 2011; Spence 2020; Spence, Wang, and Youssef 2017), a combination of herbs and spices, such as Thai curry paste, has contrasting sensory elements as well as the similarities. Imported ingredients from just a few centuries ago influenced and enabled the combination of flavors (see Chapter 2). Thanks to the country's climate and geographic location that sustain the growth of agricultural products, these ingredients (e.g., chili) have now been considered local plants.

The multisensory mixture of herbs and spices is perceived as a gestalt, in which diners cannot precisely identify each ingredient. However, if just one ingredient is missing from a Thai curry paste, the eater will no longer relate or identify the dish as it was originally intended to be. Therefore, each and every ingredient required in any curry paste (*kruang-gaeng*, เครื่องแกง) is indispensable. Thus, ensuring the availability of all ingredients is at the heart of every Thai cuisine repertoire (Kongpan 2018a, 2020; Sanitwong 1980). The innovation of components' combination and cooking techniques from the early days harmonized so well that several dishes have become national intangible heritage.

Not just taste and aroma, Thai meals always include varieties of food textures. While chewing, the types of food texture produce several different sensations. Cooked rice is soft and moist with a slight stickiness. Steamed vegetables are fibrous and moist. Fresh vegetables are crunchy and firm to the extent that an eater can hear the sound when nibbling. The clear soup is watery, while the coconut-based curry is thick and creamy. A dish with eggs such as the vegetable omelet is elastic, coarse, and fibrous. Chili dip is thick, pulpy, and lingering. The temperature of food is different too and can vary from fresh cold vegetables to hot soup.

6.5.2 Ingenuity of Thai Food Formulation

The culinary history aspect of Thai food is discussed in Chapter 2. Thailand has been a part of the Maritime Silk Roads, so Thai cuisine was influenced by multinational foreign traders. Imported ingredients like herbs and spices were successfully grown locally since the Thai climate, especially in terms of rainfalls and temperature, was close to the climate of South America and Indonesia, where many of the plant ingredients originated. With open-mindedness and keenness to play around with food, Thai people synchronized imported ingredients and cooking techniques with the locals. As a result, recipes that could have been regarded as "innovation" during the time have been ingrained and intertwined over time into national heritage. A study by Chavasit et al. (2003b) classified Thai food products into three levels of authenticity (low, medium, and high) based on their originality and the possibility of producing outside of Thailand. The products that were regarded as having the highest Thai identity are hom-ma-li rice (Thai jasmine rice) and herbs and spices paste (*nam-prig-gaeng*, น้ำพริกแกง), because they were grown, developed, and processed in Thailand only.

Thai people use the word *gloam-glom* (กลมกล่อม) to describe the deliciousness (*a-roy*, ความอร่อย) of Thai food (Kasempolkoon 2017). This term emphasizes the well-rounded, balanced, harmonious, refined, yet delicate, subtle, and smooth, almost in a humble sense, characteristics of traditional Thai food. Although particular food items can be salty, sour, sweet, and spicy, there must not be one taste or aroma that overwhelmed others. Another common description of taste in Thai language is in the phrase *preow-waan-mun-kem-ped*

(เปรี้ยว-หวาน-มัน-เค็ม-เผ็ด); the literal translation is sour-sweet-oily-salty-hot. In a data science approach to analyze authentic ingredients, from over 5900 recipes composed of cuisine from 22 nationalities, Thai cuisine is among the cuisines with the highest number of ingredients per recipe. When considering the authenticity of the ingredients (less overlapping with other national cuisines), Thai food was ranked second (behind Indian). Hence, Thai food is unique, starting from the uniqueness of ingredients to the high number of them required in each recipe. The top five authentic Thai food ingredients were identified as chili, fish sauce, garlic, lime, and coriander, respectively (Kim and Chung 2016). This finding supports the traditional concept of Thai food, i.e., the *ped* from chili, the *kem* from fish sauce, and the *preow* from lime juice. The garlic and coriander emphasize the use of herbs and spices. In addition, the finding also highlights the prominence of flavor ingredients in Thai cuisine, and not the bulk ingredients such as egg, flour, milk, meat, or vegetables like in other national cuisines.

Interestingly, from the 22 nationalities analyzed, only Thai cuisine has sourness ingredients, i.e., lime juice, among the top five authentic ingredients. Thus, even though the term "well-balanced taste" can be applied to any cuisine, only Thai food is distinguished by the high number of ingredients with the inclusion of sourness. Dishes that are not found to include sourness in other cuisines can be enhanced with sourness and gain popularity. For example, massaman curry, voted best world cuisine for many years (detail in Chapter 1), is sour from tamarind paste (*nam-ma-kam-piag*, น้ำมะขามเปียก), albeit most people may not notice the sourness as it is very well balanced with the hotness, saltiness, sweetness, creaminess of coconut milk, and aroma of herbs and spices. World-famous dishes like pad thai, *tom-yum-koong*, and *som-tum*, also feature a distinctive sharp sourness of freshly-squeezed lime juice.

In the view of foreigners, Thai food's spiciness may seem overpowering. However, for Thais, chili spiciness is needed in many dishes for well-balanced flavors (*gloam-glom*, กลมกล่อม). For Thai people in general, dishes renowned for their hot and spiciness are Southern-style food (Kongpan 2018a). In addition, all dishes served on the table as *sum-rub* are blended or harmonized (*gloam-gluen*, กลมกลืน), especially with accompanying dishes (*kruang-kiang*, เครื่องเคียง). Thus, even though many dishes contain chili, the spiciness will be toned down with other ingredients or dishes in the same meal.

Chef Chumpol Jangprai, the first chef awarded two Michelin stars for traditional Thai food, has been involved in traditional Thai cookery since early childhood with his family business. He described Thai food as follows. *Thai food is gloam-glom* (กลมกล่อม), *when cooking Thai food, I believe in 5 principles, i.e., high-quality seasoning, high-quality ingredients, wisdom and resourceful of flavor balancing, (cooking technique with) heat, and gastronomy passion.* Chef Chumpol described the Thai food flavor as an orchestra of five senses (shape, taste, smell, sound, and feel) and eight flavors (sour, sweet, creamy, salty, bitter, spicy, zesty, and astringent) (R-HAAN 2020; Sukkong 2018).

In the view of foreigners who did not grow up eating Thai food, they recognize Thai food differently. For example, a World Culinary Arts series organized by the Culinary Institute of America showcased a visit of Chef Einav Gefen to Thailand. Chef Gefen, an executive chef for Unilever Food Solutions, was guided to visit a wet market and restaurant by another famous Thai chef, M.L. Sirichalerm Svasti (aka Chef McDang). Chef Gefen described her perspective on understanding Thai food as follows. *You have to taste, taste, taste to educate your palette and to understand the complexity of balancing ingredients that are so bold in flavor and so pungent, like galangal, garlic, and shrimp paste, kaffir lime leaves, lemongrass, and chili. To put them all together is the foundation of most dishes, and get them fine-tuned to the degree where it is a harmony in your mouth. This is not something you can do just by reciting recipes from a book* (CIA ProChef 2017).

Another interesting perspective is from Chef McDang himself, who has years of experience internationally (Svasti 2015). In his perspective, Thai food foundation is the balance of saltiness, sweetness, sourness, and paste (mixture of herbs and spices) to deliver flavors. To Chef McDang, the nine distinctive ingredients commonly used are galangal, shallots, lemongrass, garlic, peppercorn, coriander root, chili pepper, and kaffir lime. Casually, he described Thai food as being strong and pungent, *it hits you like a ten-ton truck with each mouthful, you do not want to taste just the meat or the juiciness of the meat it's a mixture of taste and texture, all in one, like having a party in your mouth.*

Chef David Thompson, an Australian chef, dedicated himself to Thai cuisine and opened many Thai restaurants in many countries.

He paid attention to the culinary history perspective of Thai foods, including acquiring printed materials and taking personal lessons with local families. He authored a renowned cookbook, "Thai Food," but believes that cooking Thai food is a loose approach rather than a strict measurement from a recipe. His view regarding Thai food and seasoning characteristics is *Thai cooking is a paradox: it uses robustly flavored ingredients—garlic, shrimp paste, chilies and lemongrass—and yet when these are melded together during cooking. Often, the ingredients employed in a recipe can be an extraordinary, bewildering array of up to 20 items in a single dish. In any other cuisine, this would guarantee a cluttered and confusing finish, yet in Thai cooking, these disparate ingredients are transformed into a seamless whole—the honed result of generations of fine Thai cooks. The diverse flavors work harmoniously in concert—rounding, contrasting and supporting each other* (Thompson 2002).

All ingredients will be carefully combined within each dish, not with rigid measurement and strict cooking codes, but with talent, skills, experience, and passion. The cook will need to build their own expertise on sensory perception and evaluation, resourcefulness, and problem-solving skills to balance all the ingredients to the flavor profile of that particular dish, one saucepan at a time. Ingredients grown or fermented in different regions, seasons, or at various stages of maturity contain different flavors and chemesthetic compounds.

There is no specific principle of flavor pairing (Ahn et al. 2011) can fully describe Thai cuisine. There are similarities and contrasts in the herbs and spices paste, meat, vegetables, and even garnish ingredients. However, the food-bridging hypothesis proposed by Simas et al. (2017) may be used to explain the deliciousness of traditional Thai cuisine. *Tom-yum-koong*, for example, consists of three herbs, i.e., chili, lemongrass, and kaffir lime leaves. As seen in Figure 6.2, chili shares no common major volatile compounds with lemongrass. However, the kaffir lime leaves and chili contain limonene. Similarly, kaffir lime leaves also share one common primary volatile compound with lemongrass. Thus, the kaffir lime leaves serve as a bridge to link chili and lemongrass.

6.5.3 Multidimension of Gub-khao

The *gub-khao* with a more visible solid (little or no liquid visible on the finished dish) is identified most likely by cooking technique. For

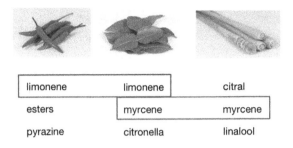

limonene	limonene	citral
esters	myrcene	myrcene
pyrazine	citronella	linalool

Figure 6.2 Kaffir lime leaves aromatic compounds served as a bridge linking chili and lemongrass in *tom-yum*.

example, *pad* (ผัด) means stir-frying. *Yum* is a spicy salad mix. *Nam-prig* is pounding and blending chili and other ingredients. Thus, the first dimension of *gub-khao* is its water content.

Any *gub-khao* that contains a lot of water is called *tom* (ต้ม) or *gaeng* (แกง). Another dominant character can further specify these soupy savory repertoires—for example, *gaeng-ped* (แกงเผ็ด) for spicy coconut milk-based curry. *Gaeng-som* (แกงส้ม) means sour soup. *Gaeng-jued* (แกงจืด) is bland clear soup.

If we consider dishes for their spiciness, they can be classified further by the proportion of herbs and spices used in the recipe (chemesthesis), especially chili. So, the use of herbs and spices seems to be a second dimension. Since some of these dishes contain fat sources such as lard, oil, and coconut milk, the third dimension is the fat content. Simple ingredients analysis based on herbs and spices (chemesthesis), fat, and water content resulted in a Gub-Khao Grid, as illustrated in Figure 6.3. The soupy *gub-khao* varieties are located on the right side of the grid (boxes 5–8). The lower boxes (3,4,7 and 8) are primarily bland, while the upper boxes are those dishes that contain herbs and spices paste.

The *nam-prig*, which is concentrated chili, herbs, and spices, is located in box 1. Solid food, like a **deep-fried spicy fish patty** (*tod-mun*, ทอดมัน), which has high chili paste concentration and high fat, is located in box 2. Other fried food, e.g., fried vegetables, fried fish, and omelet, with no herbs or spices, is found in box 3. Then those fresh, steamed, or boiled vegetable or egg or mixed salad side dish (e.g., *ar-jard*, อาจาด) is located in box 4. Likewise, for the soupy dishes matrix, the soup that requires curry paste but no coconut cream, e.g., **spicy and sour vegetable soup** (*gaeng-som*, แกงส้ม), is located in box 5. The coconut cream-based curry, which requires both curry paste and

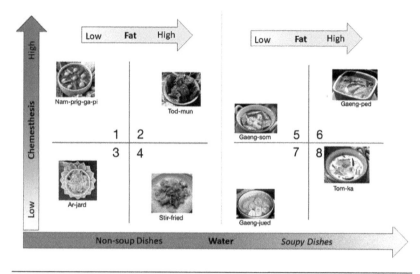

Figure 6.3 Gub-kao Grid, proposed as an illustration of all dishes in terms of their relative spiciness, water content, and fat content.

coconut cream, e.g., *gaeng-ped* (แกงเผ็ด), is positioned in box 6. Below that, in box 7, is soup with coconut milk but contains only a small amount of a few herbs or spices, such as the galangal spicy soup (*tom-ka*, ต้มข่า). Finally, clear soup with no chemesthesis such as *gaeng-jued* (แกงจืด) is located in box 8.

6.6 *Sum-rub*: The Thai Cuisine Wisdom

The Thai meal matrix is both a compliment and a contrast in each dish as well as in the whole served meal. As mentioned in Chapter 1, a Thai meal is an ensemble of many courses served all at once, called *sum-rub* (สำรับ). The main structure of the setting is rice with well-balanced taste, aroma, and texture of various savory dishes (*gub-khao*, กับข้าว) to serve along with rice. The typical component of *gub-khao* is a spicy chili dip (*nam-prig*, น้ำพริก). The chili dip is never to be served by itself; it has to be accompanied with fried fish, and with various fresh and steamed vegetables. Also in a *sum-rub*, thick soupy dish such as coconut-cream-based curry is served with its accompaniment (*kruang-kieng*, เครื่องเคียง) like clear soup that includes vegetable soup and then spicy salad or stir-fried vegetables.

6.6.1 The Accompaniment Concept (Kruang-kieng)

The concept of a supplemental or an accompanying dish is either to tone down or intensify an attribute. The primary purpose of the accompanying dish is to weaken, contrast, or counterbalance an attribute (such as color, odor, taste, texture, and oral sensation). This type of side dish is called *kruang-nam* (เครื่องแนม) or *kruang-kieng* (เครื่องเคียง). Hot and spicy curries such as red or green curry are usually served with salty meat such as fried or grilled sun-dried pork or beef. The saltiness of the meat complements the spiciness of the curry. However, some types of curry with a higher concentration of coconut cream, such as massaman curry, might leave an oily or waxy sensation afterward.

This oiliness lingering in the mouth after the food is swallowed has been described as an undesirable fat coating in the oral cavity (*lian*, เลี่ยน). To reduce this unpleasant feeling, a condiment such as vegetable pickle is a regular side dish served with curry (Maeban 2017). This type of pickle, called *ar-jard*, consists of cucumber, onion, and chili; it is sweetened with syrup, sourness is added from vinegar, and a tiny amount of salt may also be added. In other words, the dish on the right side of the gub-khao matrix (high chemesthesis, high fat, box 5) should be accompanied by a side dish that is solid, and low chemesthesis (box 3, 4, or both).

As with any other types of Thai curries, the spicy and creamy soup is eaten with the aromatic jasmine rice and the accompaniment of a specific supplementary dish (*kruang-kiang*), which is believed to enhance the deliciousness. Traditionally, recommended accompanied dish to the roasted duck red curry should be dominated by saltiness, i.e., stir-fried sun-dried salted radish with egg (*chai-po-kem-pad-kai*, ไชโป้เค็มผัดไข่), stir-fried fermented cabbage with egg (*puk-kard-dong-pud-kai*, ผักกาดดองผัดไข่), fried sun-dried salted fish, and minced pork balls (*moo-pia*, หมูเปี๊ย) (Kongpan 2018b).

Another type of *gub-khao* (main savory dish) that is always served with *kruang-kiengh*) is chili dip. One of the most popular chili dip is a shrimp paste chili dip (*nam-prik-ga-pi*, น้ำพริกกะปิ). See Figure 6.4. Compared to a Western meal course, the dip is comparable to a salad dressing as it is mainly to be eaten with a lot of vegetables and some meat. However, Thai chili dip is not as homogenous nor consistent

Box 1

Box 4   Box 3

Figure 6.4 *Ga-pi* chili dip accompanied by fried short-bodied mackerel and various vegetables. The ensemble is an example of the accompaniment concept described in terms of position (box number) on the Gub-khao Grid.

because the dip and vegetables are not meant to be mixed (always served separately). The shrimp paste chili dip is typically served with an accompaniment of fried fish, varieties of fresh and steamed vegetables, and pancake-style fried climbing wattle with egg.

There are two popular fish that accompany the chili shrimp paste dip. One is fried short-bodied mackerel (*Rastrelliger brachysoma*, known as *pla-too*, ปลาทู) and the other one is the snakeskin gourami fish (*Trichogaster pectoralis* Regan, known as *pla-salid*, ปลาสลิด) (Ninwichian et al. 2018; Senarat and Kettratad 2016; Srinulgray and Piyapattanakorn 2009). The part of the climbing wattle mixed with egg and fried is the leaves of *Acacia pennata* or *cha-om* (ชะอม) (Duncan, Chompoothong, and Burnette 2012).

In this case, the dark brownish color of the chili shrimp paste contrasts with the shades of green and the yellow color of the vegetables and the fried climbing wattle leaves with eggs. Varieties of food, especially with different colors, have been shown to increase the pleasantness of the meal (Piqueras-Fiszman and Spence 2014). The tastes of the chili dip are salty from fish sauce, sour from lime juice, sweet from palm sugar, and hot from chili peppers. There is also an umami taste from the glutamate and nucleotides in the shrimp paste (Jinap et al. 2010). The flavors of the dip are weaker when eaten with accompaniments, and the taste also varies widely between bites and different eaters throughout the meal. Texture or sensation is also an essential factor. Since each person picks a different combination, every spoonful is different.

6.6.2 The Sum-rub Table Setting Concept

In this chapter so far, we have described the multisensory properties and design principle of each savory dish (*gub-khao*). Nonetheless, another element that is the heart of Thai meal is the Thai-style table setting known as *sum-rub* (see also Chapter 1). Thai food identity is *sum-rub* meal setting, well-balanced taste, herbs and spices with potential health benefits, and beautiful presentation (Chavasit et al. 2003b). Traditionally, the *sum-rub* always includes *nam-prig* and other accompanied items. Fig 6.4 shows the *nam-prig* is in box 1, fried *pla-too* is in box 3, and the fresh and steamed vegetables in box 4.

Thai people set shared dishes of savory items (generally called *gub-khao*, กับข้าว) on a round tray. This setting dates back from Ayutthaya to Rattanakosin Period, so the tradition has lasted for several hundred years. This *sum-rub* (สำรับ) and the wisdom of this meal setting is recognized as a UNESCO national heritage. Though the details can differ among regions, they all share the same unique characteristics of serving and combination (Delwiche 2004; Kongpan 2018b; Pinket 2019; Sujachaya 2016).

When serving, an assortment of dishes (or gub-khao) is placed together in a beautiful display of all the food attributes on a large round raised tray and placed on the floor. Everyone shares the *sum-rub*, so a serving spoon is placed in every dish. Rice is served on individual plates, and the rice container is placed on the side. The setting is clearly different from the Western full course meal, as all the dishes are laid out simultaneously. Each diner picks and chooses any of the *gub-khao* with no specific pattern.

Common Thai food etiquette is simply to take a few spoons *of a gub-khao* at a time and finish one off before moving on to pick another item. Hence, the possibility is endless meaning every spoonful is always different from the previous ones (Figure 6.5). Considering the concept of sensory-specific satiety (Havermans et al. 2009; Rolls 1986), it is likely that the pleasure of the meal can be maintained for

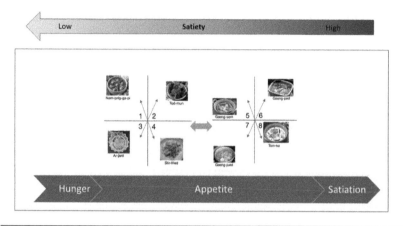

Figure 6.5 The *sum-rub* table setting concept, characterized by high meal variety with infinite combinations, and its possible relationship with satiety and satiation as the meal progresses.

longer (extended appetite stage) than if each dish needs to be finished before moving on to the next course of the meal.

The *sum-rub* custom of eating is also very versatile and contributes to the nutrient balance (Chavasit et al. 2003b). All generations can sit together and enjoy the same dishes. The young can enjoy creating their favorite menu by choosing dishes with a milder flavor, such as clear soup, fried meat, and stir-fried vegetables. The elderly can opt for softer textures such as steamed vegetables and egg dishes and compensate or customize the intensity of flavor and chemesthesis to match their preferences and sensitivity. Those who like more hot and spicy food can modify the flavor by adding side condiments like the **nam-pla-prig** (sliced chili peppers in fish sauce).

6.7 The Holistic Approach to Thai Meal Pleasantness

The experience of tasting Thai food for the first time, especially on *tom-yum-koong*, is usually described with amazement and excitement. If flavors and trigeminal senses were lights, the experience would be comparable to someone who has been in dim light for a long time. Then all of a sudden, the room is lit up with bright, blinking, flashing lights of all colors for the first time. This bold yet divine experience is a surprise and is hard to forget.

The responses to a simple question, "Please think about the first time you tried Thai food. What can you recall about that day?," are listed briefly in Appendix A. More of the similar responses are retrievable through readers' comments on many foodies and travel websites. The memory of trying Thai food for the first time is usually a vivid one, and most people will be able to describe the experience even after many years. They are likely to enthusiastically recall not just how the food tastes, smells, or feels in the mouth, but the details around that mealtime, the total pleasant experience. Similar responses can be viewed through many travel websites with comment sections from general readers. The excitement and liking toward a dish may come from the pleasant surprise of new and unexpected flavor at the complexity that is not too much (Lévy, MacRae, and Köster 2006; Spence 2020). This pleasant and memorable sensory experience may create expectation and ultimately influent future purchase or consumption behavior (Köster and Mojet 2017; Small et al. 2007).

There are both cognitive and affective drivers to hedonic responses to Thai food. While taste, quality, freshness, and appearances were important features to American diners visiting Thai restaurants in the United States (Jang, Ha, and Silkes 2009), foreign tourists visiting Thailand valued cultural and local experiences and menu atmosphere (Jeaheng and Han 2020). The emphasis on food presentation as a way to learn about culture, wisdom, and daily life also echoed in another survey on significant attributes of street food. The joyful memory of eating Thai food in the past enhances positive attitudes toward the Thai culinary tradition. The positive attitudes, which are the beliefs (food quality and taste) and emotions (fun experience and happiness), are the drivers for intention to repeat consumption as well as recommending others to acquire Thai food (Torres Chavarria and Phakdeeauksorn 2017).

There are many dimensions of Thai cuisine. In each *gub-khao* dish, palatability may come from the ingenious blend of fresh herbs and dried spices. The bridge of flavors in the combinations may enhance the deliciousness of each dish; the hotness of chili may induce benign masochism that people enjoy. The spiciness from chili and other sensations from herbs and spices create just the suitable complexity that is likable by first-timers. With more exposure, the complexity of *gub-khao* can be customized and gradually increased as the *sum-rub* style of table setting with rice allowed for an endless possibility of patterns and concentration. The fame of this signature kitchen of the world itself may become a cognitive driver attracting more people to try Thai food.

As Prof. Charles Spence (2017) put it in Chapter 9, The Meal Remembered, of his book Gastrophysics: The New Science of Eating, *no matter how good or bad a meal, it will never last more than a few hours. Our memory of a meal, at least of an enjoyable one, is where so much of the pleasure of the experience resides. It can last for days, weeks or even years. Our decision to stick with one (brand) or switch to another is often based on our recollection of the taste (of the product), or what we thought about the experience the last time we encountered it.*

6.8 Modernized Thai Cuisine

Even with an ancient recipe with all local ingredients, the traditional Thai meal can be modernized into an impressive and pleasant one. For example, Chef Chumpol Jangprai from a two Michelin-starred restaurant R-Haan in Bangkok has elaborated the wisdom of Thai cuisine with a more modern presentation. The ancient and almost unheard of snack like watermelon with dried snakehead fish (*pla-hang-tang-mo,* ปลาแห้งแตงโม) is no longer sound appetizing to anyone in the modern day, no wonder this item has almost vanished from Thai restaurant menu.

The flavor of the dish, as the ingredients imply, is an excellent example of flavor incongruity. There are the juiciness and sweetness of watermelon, the savory, earthy and salty smell of the dried fish, and topped with the sharp scent and crispiness of the deep-fried shallot. Thanks to the creativity and innovative way of presenting, this snack has been modernized to a multisensory experience that will create a long-lasting dish memory (Figure 6.6). Chef Chumpol's creation from locally sourced ingredients is intriguing, especially with the presentation that contributes significantly to visual attributes and expectations. The Northeastern red ant eggs and local leaves of the same area (*pak-wan,* ผักหวาน) were mixed into a spicy salad (Figure 6.7).

Another plant source, lotus, is typically grown commercially for its flower (Fig 6.8 a–d). Four parts of the plant include root, stem, shoot, and pollen. Though flavors are not very distinctive, they all have different textures, contrasting mouthfeel and creating different sounds when chewing. Using all parts of raw ingredients used for any dish is

Figure 6.6   Watermelon with dried snakehead fish (*plahang-tang-mo*). (Reprinted with permission from the rightful owner, R.HAAN Restaurant)

Figure 6.7   Spicy salad with red ant eggs and *pak-wan*. (Reprinted with permission from the rightful owner, R.HAAN Restaurant)

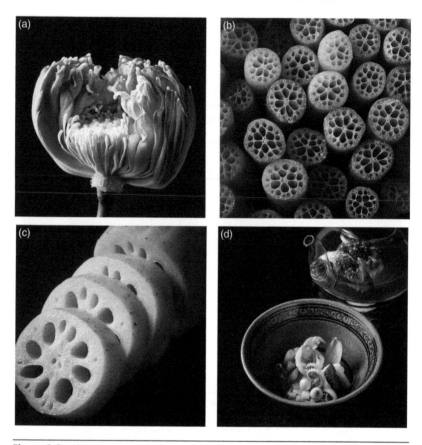

Figure 6.8   Various parts of lotus plants. (a) Pollen. (b) Stem. (c) Shoot and root. (d) As ingredients in the "Thai Sea in Ancient Soup with Ko Pha Ngan virgin coconut". A modernized from a vintage recipe. (Reprinted with permission from the rightful owner, R.HAAN Restaurant)

also a method of reducing waste. Chef Chumpol has been appointed Sustainable Food Ambassador by FeedUp@UN (Thai PBS 2021).

In the 21st century, Thailand still has a lot to explore in terms of locally sourced ingredients and meal components, waiting to be elaborated for even more theatrical, emotional, and storytelling elements and ultimately long-lasting memory of a sensory adventure and enjoyment of a meal.

6.9 Conclusion

Previous chapters have provided insights into the sensory attributes of *gub-khao* and the ingredients and cooking techniques to achieve them. Chili and the blend of herbs and spices occupied only a small proportion of a recipe but formed an identity of that particular dish.

The current chapter explored further into why the sensory properties of the food have become favorites for foreigners and a habit the
locals crave. Besides chili peppers, a comparative approach with other
national cuisines also identified authentic Thai food ingredients,
including fish sauce, garlic, lime, and coriander. The use of unique
fermented shrimp paste in herbs and spice blends is also unique. Thai
dishes also deployed a high number of ingredients, i.e., 22 on average, making the cuisine second only to India in terms of the number
of ingredients. However, the chapter also suggests that Thai cuisine's
success is not to be considered individually, but rather a design of dining set known as sum-rub contributes significantly to the Thai meal
pleasantness.

References

Ahn, Yong-Yeol, Sebastian E. Ahnert, James P. Bagrow, and Albert-László
    Barabási. 2011. "Flavor network and the principles of food pairing."
    Scientific Reports 1:196. doi: 10.1038/srep00196
Brennan, Jennifer. 1981. Original Thai Cookbook. Richard Marek publica-
    tions, New York, ISBN-:9780399510335
Byrnes, Nadia K, and John E Hayes. 2013. "Personality factors predict spicy
    food liking and intake." *Food Quality and Preference* 28 (1):213–221.
Byrnes, Nadia K, and John E Hayes. 2015. "Gender differences in the influ-
    ence of personality traits on spicy food liking and intake." *Food Quality
    and Preference* 42:12–19.
Carstens, Earl, Mirela Iodi Carstens, Jean-Marc Dessirier, Michael O'Mahony,
    Christopher T Simons, Makoto Sudo, and Satoko Sudo. 2002. "It hurts
    so good: Oral irritation by spices and carbonated drinks and the underly-
    ing neural mechanisms." *Food Quality and Preference* 13 (7–8):431–443.
Chavasit, Visith, Vongsvat Kosulwat, Somsri Charoenkiatkul, and
    Somkiat Kosulwut. 2003a. "Exporting Thai Restaurant Business."
    In *Opportunities and Potentials of Developing Thai Food to the World
    Market (การวิเคราะห์โอกาสและศักยภาพของการพัฒนาอาหารไทยสู่ตลาดโลก)*, 41–43. Bangkok,
    Thailand: Thailand Science Research and Innovation (TSRI).
Chavasit, Visith, Vongsvat Kosulwat, Somsri Charoenkiatkul, and Somkiat
    Kosulwut. 2003b. "Thai Food Identity (เอกลักษณ์อาหารไทย)." In *Opportunities
    and Potentials of Developing Thai Food to the World Market* (การวิเคราะห์โอกาส
    และศักยภาพของการพัฒนาอาหารไทยสู่ตลาดโลก), 58–60. Bangkok, Thailand: Thailand
    Science Research and Innovation (TSRI).
Choy, Mandy, Suleika El Fassi, and Jan Treur. 2021. "An adaptive network
    model for pain and pleasure through spicy food and its desensitization."
    Cognitive Systems Research 66:211–220. doi: https://doi.org/10.1016/j.
    cogsys.2020.10.006

Gefan, Einav, 2017. "Understanding Thai cuisine". In *The Culinary Institute of America's World Culinary Arts Series*. CIA videos from the Culinary Institute of America.

Cliff, Margaret, and Hildegarde Heymann. 1992. "Descriptive analysis of oral pungency 1." *Journal of Sensory Studies* 7 (4):279–290.

Cooke, Lucy, and Allison Fildes. 2011. "The impact of flavour exposure in utero and during milk feeding on food acceptance at weaning and beyond." *Appetite* 57 (3):808–811.

Dalton, Pamela, and Nadia Byrnes. 2016. Psychology of chemesthesis – Why would anyone want to be in pain? In *Chemesthesis*, 8–31.

Danhi, Robert, and Ari Slatkin. 2009. "Capturing the essence of Southeast Asia." *Prepared Foods* 178 (1):84–86.

Delwiche, Jeannine. 2004. "The impact of perceptual interactions on perceived flavor." *Food Quality and Preference* 15 (2):137–146.

Dessirier, Jean-Mark, Cas T Simons, Makoto el Sudo, Sakoto Sudo, and Earl Carstens. 2000. "Sensitization, desensitization and stimulus-induced recovery of trigeminal neuronal responses to oral capsaicin and nicotine." *Journal of Neurophysiology* 84 (4):1851–1862.

Dishman, Rod K., and Patrick J. O'Connor. 2009. "Lessons in exercise neurobiology: The case of endorphins." *Mental Health and Physical Activity* 2 (1):4–9. doi: https://doi.org/10.1016/j.mhpa.2009.01.002

Duncan, Kimberly, Niemet Chompoothong, and Rick Burnette. 2012. "Vegetable production throughout the rainy season." *Environmental Science*. https://www.semanticscholar.org/paper/Vegetable-Production-Throughout-the-Rainy-Season-Notes-Duncan/d96a07f481b653c04d697b0e6b92ed5a57e9396a

García-Estévez, Ignacio, Alba María Ramos-Pineda, and María Teresa Escribano-Bailón. 2018. "Interactions between wine phenolic compounds and human saliva in astringency perception." *Food & Function* 9 (3):1294–1309.

Haley, Howard, and Shane T. McDonald. 2016. "Spice and herb extracts with chemesthetic effects." In: *Chemesthesis: Chemical Touch in Food and Eating*, edited by S. T. McDonald, D. A. Bolliet, J. E. Hayes, Wiley-Blackwell, 32–48.

Havermans, Remco C., Tim Janssen, Janneke C. A. H. Giesen, Anne Roefs, and Anita Jansen. 2009. "Food liking, food wanting, and sensory-specific satiety." *Appetite* 52 (1):222–225.

Hayes, John E. 2016. "Types of chemesthesis I. Pungency and burn: Historical perspectives, word usage, and temporal characteristics." In: *Chemesthesis: Chemical Touch in Food and Eating*, edited by S. T. McDonald, D. A. Bolliet, J. E. Hayes, 92. https://doi.org/10.1002/9781118951620.ch6

Hutchinson, Susan E., Leslie A. Trantow, and Zata M. Vickers. 1990. "The effectiveness of common foods for reduction of capsaicin burn." *Journal of Sensory Studies* 4 (3):157–164. doi: 10.1111/j.1745-459X.1990.tb00466.x

Jang, SooCheong Shawn, Aejin Ha, and Carol A. Silkes. 2009. "Perceived attributes of Asian foods: From the perspective of the American customers." *International Journal of Hospitality Management* 28 (1):63–70.

Jeaheng, Yoksamon, and Heesup Han. 2020. "Thai street food in the fast growing global food tourism industry: Preference and behaviors of food tourists." *Journal of Hospitality and Tourism Management* 45:641–655. doi: https://doi.org/10.1016/j.jhtm.2020.11.001

2010. "Sensory attributes of dishes containing shrimp paste with different concentrations of Jinap, Selemat., Abdul. Raman Ilya-Nur, S. C. Tang, P. Hajeb, Muhammad Shahrim, and M. Khairunnisak. glutamate and 5'-nucleotides." *Appetite* 55 (2):238–244.

Jinks, Antony., and David. G. Laing. 1999. "Temporal processing reveals a mechanism for limiting the capacity of humans to analyze odor mixtures." *Cognitive Brain Research* 8 (3):311–325. doi: https://doi.org/10.1016/S0926-6410(99)00034-8

Kasempolkoon, Aphilak. 2017. "When "Royal Recipes" became more widely known: Origin and development of "Royal Cookbooks" in King Rama V's reign to girls' schools' establishment." *Vannavidas* 17:353–385. https://so06.tci-thaijo.org/index.php/VANNAVIDAS/article/view/107539

Kim, Kyung Joong, and Chang Ho. Chung. 2016. "Tell me what you eat, and I will tell you where you come from: A data science approach for global recipe data on the web." *IEEE Access* 4:8199–8211. doi: 10.1109/ACCESS.2016.2600699

Kongpan, Srisamorn. 2018a. "Identity of Thai food (อัตลักษณ์อาหารไทย)." In *Intangible Cultural Heritage food of Thailand (อาหารขึ้นทะเบียน)*, 14. Bangkok, Thailand: S.S.S.S.

Kongpan, Srisamorn. 2018b. *Intangible Cultural Heritage Foods of Thailand* (อาหารขึ้นทะเบียน มรดกทางภูมิปัญญาของชาติ). Bangkok, Thailand: S.S.S.S. (บริษัท ส.ส.ส. จำกัด).

Kongpan, Srisamorn. 2020. "Thai food that not well-known (กับข้าวไทย เรื่อง เล่าที่เราหลายๆคนไม่รู้)." In *The Story of Thai Food (เรื่องเล่ากับข้าวไทย)*, 8. Bangkok, Thailand: S.S.S.S.

Köster, Ep, and Jos Mojet. 2017. "Sensory memory." In *Time-Dependent Measures of Perception in Sensory Evaluation*, 88–123.

Laing, David G., Catherine Link, Anthony L. Jinks, and Ian Hutchinson. 2002. "The limited capacity of humans to identify the components of taste mixtures and taste–odour mixtures." *Perception* 31 (5):617–635.

Lawless, Harry, Paul Rozin, and Joel Shenker. 1985. "Effects of oral capsaicin on gustatory, olfactory and irritant sensations and flavor identification in humans who regularly or rarely consume chili pepper." *Chemical Senses* 10 (4):579–589. doi: 10.1093/chemse/10.4.579.

Lawless, Harry T., and Hildegarde Heymann. 2010. "Physiological and psychological foundations of sensory function." In *Sensory Evaluation of Food*, 19–56, Food Science Text Series. New York, NY: Springer. doi: https://doi.org/10.1007/978-1-4419-6488-5_2

Lee, Kyu-Won, and Kwang-Ok Kim. 2013. "Effects of fat and sucrose in palate cleansers on discrimination of burning sensation of capsaicin samples." *Food Science and Biotechnology* 22 (3):691–696. doi: 10.1007/s10068-013-0133-6

Lepananon, Tanarat. 2005. *Weight Management by Various Methods of Dietary Advice with Meal Replacement in the Overweight and Obese Women.* Division of Nutrition and Biochemical Medicine, Department of Medicine, Faculty of Medicine, Ramathibodi Hospital, Mahidol University, Bangkok, Thailand.

Lévy, Claudia. M., Neil A. MacRae, and Ep Köster. 2006. "Perceived stimulus complexity and food preference development." *Acta Psychologica* 123 (3):394–413. doi: https://doi.org/10.1016/j.actpsy.2006.06.006

Loss, Christopher R, and Ali Bouzari. 2016. "On food and chemesthesis – food science and culinary perspectives." In: *Chemesthesis: Chemical Touch in Food and Eating* 250.

Ludy, Mary-Jon, and Richard D Mattes. 2012. "Comparison of sensory, physiological, personality, and cultural attributes in regular spicy food users and non-users." *Appetite* 58 (1):19–27.

Maeban. 2017. Accompanied Dishes (เครื่องที่ใช้แนมกับอาหารหลัก). *Maeban.*

Marshall, Katrina, David G. Laing, Anthony L. Jinks, Juli Effendy, and Ian Hutchinson. 2005. "Perception of temporal order and the identification of components in taste mixtures." *Physiology & Behavior* 83 (5):673–681. doi: https://doi.org/10.1016/j.physbeh.2004.08.038

Namwong, Jutarat; Pewporchai, Passapong 2019. "Causative Verbs of Cooking: Study from recipe of Mom Som Cheen Rachanuprapan 1892 (คำ กริยา กา รี ต เกี่ยว กับ การ ประกอบอาหาร: ศึกษา จาก ตำรา กับ เข้า ของ หม่อม ซ่ ม จีน ราชา นุ ประพันธ์ ร. ศ. 110)." Conference Proceedings, 10th Global Goals, Local Actions: Looking Back and Moving Forward" (รายงานการประชุม วิชาการ เสนอ ผล งาน วิจัย ระดับ ชาติ และ นานาชาติ), Suan Sunandha Rajabhat University, Bangkok, Thailand.

Ninwichian, Parichart, Nirandon Phuwan, Kesara Jakpim, and Panya Sae-Lim. 2018. "Effects of tank color on the growth, stress responses, and skin color of snakeskin gourami (*Trichogaster pectoralis*)." *Aquaculture International* 26 (2):659–672.

Nitivorakarn, Saruda. 2014. "Thai food: Cultural Heritage of the Nation."*Academic Journal Phranakhon Rajabhat University* 5 (1):171–179.

Padoongpatt, Mark. 2017. "1. "One Night in Bangkok": Food and the everyday life of empire." In *Flavors of Empire*, 24–55. University of California Press, California, USA.

Pewporchai, Passapong. 2017. "A study of cooking terms in Thai recipe book: A case of her ladyship plain Phassakorawong's "Mae Krua Hua Pa" recipe book." *Journal of Liberal Arts, Ubon Ratchathani University* 13 (2):138–165.

Pinket, Piset. 2019. "Thai setting tray of food since Ayutthaya to Rattanakosin period (การ จัด สำรับ อาหาร ของ คน ไทย สมัย อยุธยา จนถึง สมัย รัตนโกสินทร์)." *Journal of Ayutthaya Studies Institute* 11 (1):32–42.

Piqueras-Fiszman, Betina, and Charles Spence. 2014. "Colour, pleasantness, and consumption behaviour within a meal." *Appetite* 75:165–172. doi: https://doi.org/10.1016/j.appet.2014.01.004

Plainoi, Sor. 2015. *Krua Thai (Thai Kitchen).* Bangkok, Thailand: Sarakadee.

R-HAAN. 2020. "R-HAAN Offers Authentic Thai Food in Bangkok in Their Summer Menu." R-HAAN. https://www.r-haan.com/en/summer-samrub-2020.

Rolls, Barbara J. 1986. "Sensory-specific satiety." *Nutrition reviews (USA)*.

Rozin, Paul, and Deborah Schiller. 1980. "The nature and acquisition of a preference for chili pepper by humans." *Motivation and Emotion* 4 (1):77–101.

Samant, Shilpa S., Sungeun Cho, Andrew D. Whitmore, Syllas B. S. Oliveira, Thais B. Mariz, and Han-Seok Seo. 2016. "The influence of beverages on residual spiciness elicited by eating spicy chicken meat: Time-intensity analysis." *International Journal of Food Science & Technology* 51 (11):2406–2415.

Sanitwong, Mom Ractchawongse Tuang 1980. "The important things to know (สิ่งสำคัญที่ควรทราบ)." In *Tumrub Sai Yaowapa (ตำรับสายเยาวภา)*, edited by Yaovabha Bongsanid, 34–35. Bangkok, Thailand: Saipunya Samakom.

Schaal, Benoist, and Karine Durand. 2012. "The role of olfaction in human multisensory development." In: *Multisensory Development*, edited by Andrew J. Bremner, David J. Lewkowicz, and Charles Spence. 2012, ISBN-13: 9780199586059. doi: 10.1093/acprof:oso/9780199586059.001.0001

Schaal, Benoist, Luc Marlier, and Robert Soussignan. 2000. "Human foetuses learn odours from their pregnant mother's diet." *Chemical Senses* 25 (6):729–737. doi: 10.1093/chemse/25.6.729

Schmidt, Darlene. 2021. A Guide to Thai Food and Culture. doi:https://www.thespruceeats.com/thai-food-and-culture-3217393#thai-cutlery-and-eating-stylehttps://www.thespruceeats.com/thai-food-and-culture-3217393#thai-cutlery-and-eating-style.

Senarat, Sinlapachai, and Jes Kettratad. 2016. *Structure and Alteration of Gonadotropin Releasing Hormone – 1 Peptidergic Neuronal System During Breeding Season of Short mackerel Rastrelliger brachysoma (Bleeker, 1851) from Samut Songkhram Province.* Chulalongkorn University.

Seubsman, Sam-ang, Pangsap Suttinan, Jane Dixon, and Cathy Banwell. 2009. "20 - Thai meals." In *Meals in Science and Practice*, edited by Herbert L. Meiselman, 413–451. Cambridge: Woodhead Publishing.

Simas, Tiago, Michal Ficek, Albert Diaz-Guilera, Pere Obrador, and Pablo R. Rodriguez. 2017. "Food-bridging: A new network construction to unveil the principles of cooking." *Frontiers in ICT* 4 (14). doi: 10.3389/fict.2017.00014

Singsomboon, Termsak. 2015. "The use of Thai food knowledge as marketing strategies for tourism promotion." *Thammasat Review* 18 (1):82–98.

Small, Dana M., Genevieve Bender, Maria G. Veldhuizen, Kristin Rudenga, Danielle Nachtigal, and Jennifer Felsted. 2007. "The role of the human orbitofrontal cortex in taste and flavor processing." *Annals of the New York Academy of Sciences* 1121 (1):136–151. doi: https://doi.org/10.1196/annals.1401.002

Smutzer, Gregory, Jeswin C. Jacob, Joseph T. Tran, Darshan I. Shah, Shilpa Gambhir, Roni K. Devassy, Eric B. Tran, Brian T. Hoang, and Joseph F. McCune. 2018. "Detection and modulation of capsaicin perception in the human oral cavity." *Physiology & Behavior* 194:120–131. doi: https://doi.org/10.1016/j.physbeh.2018.05.004

Spence, Charles, Qian Janice Wang, and Jozef Youssef. 2017. "Pairing flavours and the temporal order of tasting." *Flavour* 6 (1):4. doi: 10.1186/s13411-017-0053-0.

Spence, Charles. 2017. *Gastrophysics: The New Science of Eating.* UK: Penguin.

Spence, Charles. 2018. "Why is piquant/spicy food so popular?" *International Journal of Gastronomy and Food Science* 12:16–21. doi: https://doi.org/10.1016/j.ijgfs.2018.04.002

Spence, Charles. 2020. "Chapter 10 – Multisensory flavor perception: A cognitive neuroscience perspective." In *Multisensory Perception*, edited by K. Sathian and V. S. Ramachandran, 221–237. Academic Press.

Spinelli, Sara, Alessandra De Toffoli, Caterina Dinnella, Monica Laureati, Ella Pagliarini, Alessandra Bendini, Ada Braghieri, Tullia Gallina Toschi, Fiorella Sinesio, Luisa Torri, Flavia Gasperi, Isabella Endrizzi, Massimiliano Magli, Monica Borgogno, Riccardo di Salvo, Saida Favotto, John Prescott, and Erminio Monteleone. 2018. "Personality traits and gender influence liking and choice of food pungency." *Food Quality and Preference* 66:113–126. doi: https://doi.org/10.1016/j.foodqual.2018.01.014

Sprouse-Blum, Adam S., Greg Smith, Daniel Sugai, and F Don Parsa. 2010. "Understanding endorphins and their importance in pain management." *Hawaii Medical Journal* 69 (3):70.

Srinivasan, Krishnapura. 2016. "Biological activities of red pepper (*Capsicum annuum*) and its pungent principle capsaicin: A review." *Critical Reviews in Food Science and Nutrition* 56 (9):1488–1500. doi: 10.1080/10408398.2013.772090

Srinulgray, Theerarak, and Sanit Piyapattanakorn. 2009. "Genetic diversity of short mackerel *Rastrelliger brachysoma* populations in the Gulf of Thailand and Andaman Sea revealed by ISSR marker." 47. Kasetsart University Annual Conference, Bangkok (Thailand), 17–20 Mar 2009.

Stevenson, Richard J, and John Prescott. 1994. "The effects of prior experience with capsaicin on ratings of its burn." *Chemical Senses* 19 (6):651–656.

Sujachaya, Sukanya 2016. "Sumrub Thai (สำรับไทย)." In *Knowledge and Practices in Nature and the Universe: Intangible Cultural Heritage (*ความรู้และแนวปฏิบัติเกี่ยวกับธรรมชาติและจักรวาล: มรดกภูมิปัญญาทางวัฒนธรรมของชาติ*)*, 32–33. Bangkok, Thailand: Department of Cultural Promotion.

Sukkong, Ploychan. 2018. "The Heart of Thai Cuisine (ถึงแก่นสำรับไทย เจนจัดในรสวัตถุดิบ ณ อาหาร (R.HAAN) ร้านอาหารไทยที่นำบริบทไทยๆ มาใส่ในทุกอณู)." *The Standard*, 18 April.

Sukphisit, Suthon. 2019. "Chilli's Complicated History; Where did this most indispensable Thai cooking spice originate?" *Bangkok Post*, 5 May 2019, B Magazine. https://www.bangkokpost.com/life/social-and-lifestyle/1672304/chillis-complicated-history.

Svasti, Sirichalerm. 2015. Thai Food – Green Curry with Chef McDang. edited by wocomoCOOK: wocomoCOOK.

Terasaki, Masaharu, and Sumio Imada. 1988. "Sensation seeking and food preferences." *Personality and Individual Differences* 9 (1):87–93. doi: https://doi.org/10.1016/0191-8869(88)90033-5

Thai PBS. 2021. " "Thai food is the most delicious medicine" – Chef Chumpol." https://www.thaipbsworld.com/thai-food-is-the-most-delicious-medicine-chef-chumpol/

Thompson, David. 2002. "Characteristics of Thai food and seasoning." In *Thai Food*, 133–134. London, UK: Pavilion Books.

Torres Chavarria, Luis Carlos, and Panuwat Phakdee-auksorn. 2017. "Understanding international tourists' attitudes towards street food in Phuket, Thailand." *Tourism Management Perspectives* 21:66–73. doi: https://doi.org/10.1016/j.tmp.2016.11.005

Touska, Filip, Lenka Marsakova, Jan Teisinger, and Viktorie Vlachova. 2011. "A "cute" desensitization of TRPV1." *Current Pharmaceutical Biotechnology* 12 (1):122–129.

Westerterp-Plantenga, Margriet. S., Smeets, Ann and Manuela P. G. Lejeune. 2005. "Sensory and gastrointestinal satiety effects of capsaicin on food intake." *International Journal of Obesity* 29 (6):682–688. doi: 10.1038/sj.ijo.0802862

Wilson, Marie M. 1965. *Siamese Cookery*. Rutland, Vt.: C.E. Tuttle Co.

Xue, Yong, Tingchao He, Kai Yu, Ai Zhao, Wei Zheng, Yumei Zhang, and Baoli Zhu 2017. "Association between spicy food consumption and lipid profiles in adults: A nationwide population-based study." *British Journal of Nutrition* 118 (2):144–153. doi: 10.1017/S000711451700157X

7

GUB-KHAO RECIPES

From Past to Present, from Arts to Science

VALEERATANA K. SINSAWASDI
AND NITHIYA RATTANAPANONE

Contents

DOI: 10.1201/9781003182924-10

171

7.1 Introduction

The flavor attributes of most *gub-khao* dishes lie in the taste, aroma, and trigeminal sensations. The right combination and proportion of seasoning, herbs, and spices are crucial to each dish's identity. Traditionally, meat is less influential, and land animal meat was not a significant part of Thai people's diet. Considering Thailand's tropical climate, where temperatures rarely fall below 25°C during the day, it was challenging to commercialize raw meat without risks of spoilage and foodborne illnesses. Therefore, the slaughter of a four-legged animal for consumption was infrequent and reserved only for large crowd gathering occasions or celebrations. The abundance of herbs and spices used in the cuisine also helped mask any foul odors and extend the shelf life of the food. The meat ingredient, i.e., beef, pork, chicken, meat cuts with bones, offals, freshwater prawns, and fish, are mostly interchangeable in a recipe. Before electricity, seafood was consumed mainly in a preserved form such as sun-dried, salted, or fermented. The proximity to the shoreline caused the seafood to perish before reaching the city households. Accordingly, seafood was not a common classical ingredient.

The mouthfeel of meat, tenderness, and juiciness, is not so much a dominant feature in Thai cuisine as it is in other national cuisines. Conversely, the appearance, texture, and sound attributes of plant ingredients are typically crucial. Thanks to the plant diversity, there are always a wide variety of vegetables to choose from. Within a short walking distance, most suburban households have a variety of edible plants to complement any meals, either naturally grown or homegrown. The niche, non-commercial, native plant food is, for example, leave shoots of a mango tree, young jackfruit, mushrooms, and flowers. In combination, they give not just varieties of mouthfeel, e.g., crispness and crunchiness, but also sound. Different plant parts offer a delicate balance of textures, such as firmness, chewiness, toughness, fibrous, and softness. Cooking techniques also help soften the pectin and cellulose in the plant cell wall to the desired consistency. Many dishes in the recipes do not specify the type or amount of vegetables. Frequently, the phrase "any vegetables to your liking" (ผักต่าง ๆ ตามชอบ) is indicative of the versatility of the ingredients, the ingenuity of most home cooks, and the endless possible sensory properties of each *gub-khao*.

Apart from the mixture of solid and liquid ingredients (e.g., chili dip, stir-fried, clear soup), *gub-khao* can be a colloidal system. There is an emulsion, e.g., curry with coconut milk, and gel, e.g., steamed egg (*kai-toon*, ไข่ตุ๋น) and steamed fish curry (*hor-mhok*, ห่อหมก). The formation of Thai meals can be entranced by sophisticated flavor and artistic presentation, like in the case of the royal court cuisine, or dramatic and adventurous style like street foods.

7.1.1 Classic Thai Kitchen

In a traditional Thai kitchen, clay was the primary material for making both stoves and cookware. The cookstove fueled with wood or charcoal was originally called *tao-cherng-gran* (เตาเชิงกราน), then the Chinese brazier (*tao-ung-lo*, เตาอั้งโล่) had become more popular. For cookware, the most common material was also clay. A combination of charcoal and heated unglazed terracotta created a natural smoky aroma of charcoal and the earthy aroma of clay. Rice cooked in this manner is praised for its unique aroma.

When gas hob became available, temperature control was much more manageable. The metal cookware allows faster conduction heat transfer. Then, the refined cooking oil with a lower heat specific capacity and a higher boiling point cooks the food faster and facilitates changes in food color and flavor due to the Maillard browning reaction. Shorter heating time also enables different texture development, such as the contrast between the crispy surface and the moistness of the middle.

For sticky rice or glutinous rice, steaming in a woven bamboo basket called *huad* (หวด) is still popular today. The freshly cooked sticky rice will be spread out into a thin layer to let the steam escape freely. The technique is called *sai-khao* (สายข้าว), similar purpose to fluffing the white rice (see Chapter 4, topic 4.2.3). The cooled sticky rice should be somewhat cohesive during chewing but not gluey or mushy to the touch.

Mortar and pestle were another Thai kitchen essential. Popular dishes traditionally requiring their use are chili dip (*nam-prig*, น้ำพริก) and curry paste (*nam-prig-gaeng*, น้ำพริกแกง). Preparation with this technique is labor-intensive because the cook needs to lift the heavy pestle hundreds of times to crush the ingredients. The motion is not a simple repetitive action, but skill is required to achieve the desired consistency and character. Nevertheless, many home cooks still swear by the use of traditional crushing of herbs and spices with mortar and pestle for curry paste (evidenced by discussion of this topic as appeared on many websites and social media). They claim that the paste texture has the right thickness. The curry feels smooth and creamy when added to the coconut soup base. However, the curry paste prepared from an electric blender seems to have a runny consistency, and the finished dish lacks creaminess. The blade with the blender's shear force may have shortened the long chain of non-starch polysaccharides in herbs and spices, so the viscosity is lower than the paste from traditional pounding. So far no scientific consensus on this controversy.

Many Thais love the traditional ways for the slow development of flavor. With each pounding, different volatile compounds are released. Especially important when making chili dip, an experienced cook can use the aroma to guide which, when, and how many ingredients should follow. For example, pounding garlic cloves, with more cell damage, aroma precursor compounds in the cytoplasm part of the cell will contact enzymes released from the vacuole, giving a distinctive aroma to garlic. As the pounding continues, the cook will estimate if enough garlic is crushed or more is needed. The same scenario applies to adding chili pepper, shrimp paste, lime juice, palm sugar, fish sauce, etc. Hence, many home cooks still prefer to rely on their sense of taste and smell to orchestrate ingredients and techniques into one single dish.

The traditional way of pounding the pestle also allows selectively to crush ingredients to various sizes and choose which should be left in larger pieces as homogeneity is not always desirable for chili dip. The slow harmony of the pestle crushing and the adding of ingredients is theatrical to watch and results in finished products with a texture and flavor that a modern electrical appliance cannot always completely facilitate. Nonetheless, many households no longer have a mortar and pestle, once a symbol of Thai cuisine, because of lifestyle changes.

7.1.2 Classic Thai Cookbooks

General ingredients and procedures of selected dishes are detailed in this chapter. The recipes mentioned in this book are selected for various reasons, as per the details below.

1. Their recognition worldwide and also honored as Thai national heritage: *tom-yum-koong* (8.2), green curry (8.3), massaman curry (8.4), and *som-tum* (8.5).

2. The common in everyday Thai home-cooking techniques: *nam-prig-ga-pi* (8.6), vegetable stir-fry (8.7), *gaeng-lieng* (8.8), red curry (8.9), and spicy salad (8.10).

3. Demonstration of fermented ingredients with possible umami and kokumi compounds, i.e., *nam-prig-pla-ra* (8.11).

Notable Thai cuisine cookbooks, except Ninrat's (1996), were mentioned in Chapter 2. For this chapter, the author's team re-created selected vintage recipes, mostly from the *Mae-Krua-Hua-Pa* cookbook. The ingredients, recipes, cooking techniques, and sensory properties of the finished products from MKHP compared with other modern cookbooks were assessed for modification over time and perhaps the evolution of Thai cuisine.

The list of cookbooks frequently used as references is as follows:

1. ***Mae-Krua-Hua-Pa***, written by Thanphuying Plian Phassakorawong (Phassakorawong 1908); Lady Chef Cookbook, ตำราแม่ครัวหัวป่าก์, ท่านผู้หญิงเปลี่ยน ภาสกรวงศ์.

   The complete cookbook, composed of five volumes, has been published in Vajrayana Digital library's public domain (https://vajirayana.org/). Printed copies were also available commercially.

   The book will be referred to as **MKHP** in this chapter.

2. ***Tumra-Gub-Khao***, written by Mom Somjeen Rachanuprapan (Rachanuprapan 1890); Savory Dish Cookbook, ตำรากับเข้า, หม่อม ซ่มจีน ราชานุประพันธ์ the book will be referred to as **TGK** in this chapter.

3. ***Tumrub Sai Yaovabha***, written by Princess Yaovabha Bongsanid (Bongsanid 1935); Yaovabha Family Cookbook, ตำรับสายเยาวภา, พระเจ้าบรมวงศ์เธอ พระองค์เจ้าเยาวภาพงษ์สนิท.

   The book will be referred to as **TSY** in this chapter.

4. ***Cheewit-Nai-Wung***, written by M.L. Nuang Ninrat (Ninrat 1996); Life in the Palace, ชีวิตในวัง, ม.ล. เนื่อง นิลรัตน์.

   M.L. Nuang lived with her paternal grandmother, who was the chief of the royal court kitchen. The book was launched in 1996 as a collection of her magazine articles first written in 1985. The details and recipes came from her early life experience, spending her childhood in the royal palace around 1915–1935. This book is very famous and highly sought-after because of the exclusive and coveted lifestyle in the old days' royal court. Although the book is more of an autobiography than a cookbook, it includes many authentically royal court cuisine recipes (*ar-harn-chow-wung*, อาหารชาววัง). In addition, M.L. Nuang wrote the book later in her life, so the language, name of ingredients, and measurement are contemporary-style and easy to follow.

The book will be referred to as **CNW** in this chapter.

5. *Turmra-Gub-Khao-Sorn-Loog-Larn* written by Thanphuying Kleeb Mahidhorn (Mahidhorn 1949); Cookbook to Teach Offsprings, ตำรากับข้าวสอนลูกหลาน, ท่านผู้หญิง กลีบ มหิธร.

The book has been published in Vajrayana Digital library's public domain (https://vajirayana.org/). Printed copies were also available commercially.

The book will be referred to as **TGKSLL** in this chapter.

6. *Rattanakosin* Dishes 1982 written by M.L. Terb Xoomsai (Xoomsai 1982) กับข้าวรัตนโกสินทร์ 2525, หม่อมหลวง เติบ ชุมสาย.

The book will be referred to as **RD1982** in this chapter.

7. *National Heritage Food of Thailand* (Kongpan 2018), อาหารขึ้นทะเบียน มรดกภูมิปัญญาทางวัฒนธรรมของชาติ, ศรีสมร คงพันธุ์.

The book will be referred to as **NHFT** in this chapter.

The re-created dishes of MKHP recipe were compared with the contemporary recipe to explore the possible development of each dish. For authenticity of the classic dishes, the definition from the Intangible Cultural Heritage, Department of Cultural Promotion, Ministry of Culture is used as a guideline. Contemporary recipes are listed here without specifying ingredient measurement as it is beyond the scope of this book. However, more precise and duplicatable recipes are widely available on the internet and as printed materials. Recommended open resources include the detailed recipes written in Thai and English are on the *Recipes of Thai Food* web page (http://www.thaifoodheritage.com/en/recipe_category), the *Authenticity of Thai Food* web page (http://authentic.nfi.or.th), both provided by the National Food Institute, Thailand Ministry of Industry, and the *Thai SELECT* website (https://www.thaiselect.com/en/recipes) administered by the Ministry of Commerce, the Royal Thai Government.

7.2 Sour Soup with Shrimps and *Tom-yum-koong*

7.2.1 *Vintage Recipes*

The term "*tom-yum*" appeared in the first two Thai cookbooks, MKHP and TGK.

7.2.1.1 *MKHP Recipe for Out-of-the-Pot Soup or Tom-yum-khmer*   (Fig. 7.1)
This recipe from the MKHP cookbook has two names. First, out-of-pot soup (*gaeng-nog-mhor*, แกงนอกหม้อ), the name hints a unique procedure of pouring boiling stock over prepared pieces of other ingredients in a bowl (as opposed to boiling everything in a saucepan). The second name is *tom-yum-khmer* (*tom-yum-ka-men*, ต้มยำเขมร), a literal translation is Cambodian-style spicy soup. However, there was no further explanation of how the name was obtained.

   Ingredients:

- Solid ingredients: Dried snakehead fish snakehead fish (*pla-chon*, ปลาช่อน), grilled snakeskin gourami fish (*pla-salid*, ปลาสลิด),

Figure 7.1   MKHP out-of-the-pot soup, or *tom-yum-ka-men*.

fresh water prawns, sour green mango, cucumber, pickled garlic, spur chili (long red chili), coriander leaves
- Soup ingredients: whole fresh water prawns, fish sauce or salt
- Seasoning: garlic pickle juice, lime juice, sugar.

Procedures: This soup is to be prepared in 3 main steps.

- Preparation of prawn stock: Add prawns to boiling water and season with fish sauce or salt, then remove the prawns from the water.
- Preparation of ingredients: Cut and slice the cooked prawns, dried snakehead fish, grilled snakeskin gourami, sour green mango, cucumber, and pickled garlic cloves to preferred size and proportion.
- To serve: Line soup bowls with all the sliced ingredients, then pour the boiling prawn stock over the solid materials—season with lime juice, pickled garlic juice, and sugar as desired. Garnish with red spur chili slices and coriander leaves.

7.2.1.2 TGK Recipe for Tom-yum There were many varieties of the *tom-yum* (spicy soup, ต้มยำ), for example, snakehead fish *tom-yum* (*tom-tum-pla-chon*, ต้มยำปลาช่อน), termite mushroom *tom-yum* (*tom-yum-hed-kone*, ต้มยำเห็ดโคน), swamp eel *tom-yum* (*tom-yum-pla-lai*, ต้มยำปลาไหล), and chicken *tom-yum* (*tom-yum-gai*, ต้มยำไก่). There was no stock preparation. All recipes were water-based, except the chicken *tom-yum* that includes coconut milk. Ingredients of these spicy soup recipes bear very little resemble of the today's *tom-yum-koong*. These ingredients are uncooked rice (*khao-sarn*, ข้าวสาร), lemongrass, roasted dried-chili, shallot, garlic, salt, krill paste, lime juice, roasted chili paste (*nam-prig-pao*, น้ำพริกเผา), pickled garlic (*gra-tium-dong*, กระเทียมดอง), sour green mango, roasted garlic, and coriander leaves to garnish. The chicken *tom-yum* is unique among other *tom-yum* recipes in this cookbook because it requires coconut milk, coriander root, and coriander seed.

7.2.2 Contemporary Tom-yum-koong (Fig. 7.2)

Ingredients: Shrimps or prawns; water; lemongrass, thinly sliced; fish sauce; lime juice; kaffir lime leaves, torn into large stripes on both sides of the vein (but the leave still intact); bird's eye chili, preferably green to contrast with the orange-colored soup; coriander leaves; fish sauce; lime juice to season.

Figure 7.2 Contemporary *tom-yum-koong*.

Instruction: Prepare the shrimps (wash, peel, remove head and tail, devein). Squeeze and reserve the fatty mass at the shrimps' heads. Prepare shrimp stock by adding heads and tails to boiling water and boil until the peel's red hue developed. Discard the peel and head. Strain the stock if the clear soup is desired. Boil the stock, add lemongrass and fish sauce. Add shrimps, kaffir lime leaves, and chili. Season in a serving bowl with freshly squeezed lime juice. Garnish with whole chili and coriander leaves.

7.2.3 Culinary Aspects

There was no *tom-yum* spicy soup in any five volumes of the MKHP. The closest resemblance of *tom-yum-koong* is the out-of-pot soup (*gaeng-nog-mhor*, แกงนอกหม้อ) or Khmer curry (*gaeng-ka-men*, แกงเขมร). The soup base is shrimp stock and seasoned with the sourness of green mango and lime juice, saltiness from fish sauce, and spiciness from chili. The assembly without boiling keeps the cucumber and the fried gourami crisp and fresh. Dried fish and grilled gourami will add flavor dimension to the dish. The dish will be made to have a dominantly sour and salty taste with a little hint of sweetness undertone.

The re-creation outcome of this recipe impressed the author's team. There are no herbs or spices in this dish, so the aroma is very mild. The MKHP recipe did not suggest the amount of each ingredient. However, to everyone's (in the kitchen lab) surprise, the soup was very delicious and refreshing. Of all the MKHP recipes we re-created, this soup was the most memorable; our team cannot think of any exact replacement with a contemporary dish we are all accustomed to. The aroma of gourami and pickled garlic was strong but almost non-detectable in the finished product. The contrasting texture of crispiness and softness is just right. Though the recipe was called *tom-yum*, it barely resembles today's *tom-yum-koong*, everyone knows. Albeit the *gaeng-nog-mhor* is unheard of for most Thai people, the soup is not typically offered commercially anymore.

Fish sauce is a crucial seasoning for this dish. Just like all dishes without herbs and spices, such as stir-fry and clear soup, the aroma from fish sauce dominates the dish's characters. Thus, the selection of fish sauce is crucial. Lower grade fish sauce is commonly a diluted version of pure fish sauce. This cheaper version of the fish sauce may have been enhanced with food additives like MSG and color, but the characteristic aroma of good fish sauce from volatile fatty acids, etc., is lacking. Another well-known technique is to add fish sauce while the soup is hard-boiling. The rule of thumb is always to add fish sauce when the food is really hot. Adding fish sauce to the soup before boiling or to the cold vegetables in the stir-frying pan will result in an off-flavor of the dish. Volatile compounds of fish sauce were identified in pure raw fish sauce, not in dishes. Further investigation on the fish sauce aroma with different cooking techniques is needed to explain the difference in the aroma profile.

The *tom-yum-koong* was invented just recently, as detailed in Chapter 2. The combination of lemongrass and kaffir lime leaves imparts the distinctive signature aroma of the dish. Controversially, many of the recipes suggest the addition of galangal. However, for many, the addition of the galangal overpowers other ingredients. In addition, if there is galangal and coconut milk, the dish is called *tom-kha* (ต้มข่า).

7.2.4 Science Aspects

The *tom-yum-koong* is a representative of Thai cuisine in the eye of foreigners. The name implies that it is soupy, with a hot and sour taste. When heated, protein in shrimps becomes denatured, causing the shrimp body to curl up. Some varieties of shrimps and prawns contain collagen that can be converted to gelatin with moist heat giving a more elastic texture to cooked shrimps.

In general, shrimps should be the last ingredients to add to the boiling soup because its protein is easily denatured. The longer the shrimp is boiled, the higher the degree of denaturation, resulting in an overly tough and dry texture. In addition, undesirable shrinkage of the shrimps occurs, especially when the inner temperature reaches above 70°C. In addition, upon heating, the color of shrimps will become more orange. This color change results from a type of orange carotenoid pigment, astaxanthin, which is being released from binding with protein. In raw shrimp, the binding of carotenoid and protein or carotenoprotein masks the orange color of astaxanthin (Benjakul et al. 2008; Mizuta et al. 1999).

Inedible parts such as shells, gills, claws, and heads provide a pleasant shrimp aroma when cooked, so they are used to make stock and then discarded. Protein content in shrimp soup made from shrimp shells and heads was reported at over 5 g/100 g. The soup also contains around 650 mg/100 g of glutamic acid. Considering that typical *tom-yum-koong* has mushroom and fish sauce, which also provides glutamic acid, the total glutamic acid in the soup is enough to provide umami taste without added MSG. The *tom-yum-koong* can be prepared just with the shrimp stock (clear soup or *tom-yum-nam-sai*) or with the addition of a fat source (creamy soup or *tom-yum-nam-kon*). Many recipes call for *nam-prig-pao* or roasted chili paste. This product will give the aroma of roasted herbs and chili along with orange color oil. In this recipe, the orange pigment in oil drops comes from chili carotenoids rather than from the shrimp shells.

Some Thais refuse the concept of adding fat sources (milk, evaporated milk, or coconut milk) and the *nam-prik-pao* claiming that it is not authentic. They believe *tom-yum-koong* should not be milky one with orange-colored oil floating on top. In other words, there is no such dish as the so-called **tom-yum-koong-nam-kon** (creamy *tom-yum-koong*). Sensory evaluation on the different recipes of *tom-yum-koong* showed that shrimp shells stock was preferred over chicken bone stock on the *tom-yum* flavor attribute. When evaluated against the soup with added fat ingredients, the evaporated milk was preferred over coconut milk. However, when comparing the best of clear *tom-yum* and milky *tom-yum*, there were no significant differences in overall liking (Suwankanit et al. 2015). Considering that the first Thai cookbook just 150 years ago did not have any *tom-yum-koong* recipe, it is clear that Thai foods evolved and continue to evolve.

Shrimp fat (Fig. 7.3) found around the head part provides a desirable shrimp aroma and visible orange oil droplets floating on the bowl of soup. The orange pigment, astaxanthin, is soluble in fat, so the color is concentrated in the oil. The *tom-yum-koong* prepared from frozen peeled shrimps, containing only white shrimp meat, will lack these characteristic flavors, juiciness, and visual cues. Shrimp has hepatopancreas, a gland near the head that releases proteolytic enzymes to the rest of the body after death. The enzyme degrades the protein of the tail muscle and causes the shrimp texture to become undesirably soft and mushy. Thus, deheading soon after death is a necessary processing step for frozen shrimps to maintain a firm texture when thawed and cooked (Erickson et al. 2007; Sriket, Benjakul, and Visessanguan 2011).

Figure 7.3 Shrimp fat.

Originally *tom-yum-koong* is a clear soup. Depending on the preparation of shrimp stock, as discussed, if shrimp heads with high-fat content are used, the soup may contain high fat. If not, the soup is likely to be mostly water with little fat to extract and absorb aroma compounds, especially essential oil releasing from the lemongrass and kaffir lime leaves, so it is better to be prepared fresh just to the quantity needed. Leftover *tom-yum-koong* will have little flavor described as flat (ชืด). In addition, upon reheating, the texture of shrimp will be harder and dryer. Thus, the dish may be prepared ahead of time by separating all main ingredients, with shrimps just blanched to avoid dryness.

Pungency of hotness can be adjusted with how much chili is used and how it is prepared. The bird's eyes chili, which is called Thai chili by some foreigners, is very hot. The degree of hotness depends on the pungent compounds, capsaicinoids (mainly capsaicinoids and dihydrocapsaicin). The capsaicinoids are neither stored in the flesh nor the seeds but stored in little droplets in the white placenta in the middle of the fruit (Kethom, Tongyoo, and Mongkolporn 2019). Thus, the degree of pungency and hotness can be increased simply by damaging or bruising the chili fruit to release more capsaicinoids. The hotness is as rated by the Scoville scale in the unit of Scoville Heat Unit (SHU), and the bird's eye chili is among the highest compared to other variety.

The soup is best when it is seasoned and customized to individual preferences. The freshness of essential oil from the lime rind and the acidity from the lime juice are well balanced with saltiness, umami, the aroma of fish sauce, and chili's burning sensation.

7.2.5 Modification and Variation

Some recipes replace shrimp stock with chicken stock, but the flavor will be different.

Lemongrass is edible, especially when thinly sliced. But if the strong aroma and fibrous mouthfeel are undesirable, a whole stalk of the lemongrass can be used. In order to extract as much essential oil from the tight role leaves, pound the stalk with either a side of a big knife or a pestle to create bruises and cuts. The long stalk can be tied into a knot to fit in a saucepan and increase the surface area for the essential oil to leach out.

Fresh kaffir lime leaves have oil glands that are more accessible since the leaf is very thin and oil glands are located just on the surface of the leaves. Hand crushing and tearing of the leave are enough to release the essential oil. Since the aromatic oil is highly volatile, do not boil them for too long.

To increase the hotness and spiciness from chili, pound or bruise the chili pepper to expose more surface area.

When lime juice is added, the soup will be very aromatic. The juice's acidity will further denature the protein in the soup. As a result, the clumps of solid mass may be visible throughout the soup.

The reserved shrimp fat can be heated separately with cooking oil and added to the soup after shrimps.

A creamy type of soup can be prepared by adding coconut milk, usually accompanied by fried chili paste (*nam-prig-pao*, น้ำพริกเผา).

Reduce heat just to simmer to avoid the tough and dry texture of the shrimps. Serve immediately.

If other meats are used, add galangal with the lemongrass.

Mushrooms can be added to increase nutritional value.

7.3 Green Curry

Green curry (*gaeng-keow-wan*, แกงเขียวหวาน): Literal translation for this dish is sweet and green curry. However, the dish is spicy from chili, salty, and slightly sweet.

7.3.1 MKHP Recipe (Fig. 7.4 a–e)

On MKHP, there was no full recipe detailed for green curry. Instead, the author simply cited a poem describing a general chicken curry and then mentioned the modification.

Basic curry paste ingredients: Dried chili, lemongrass, fermented shrimp paste, salt, kaffir lime zest, peppercorn, galangal, coriander seed, cumin, garlic.

Modification required to make green curry paste: Replace dried chili with fresh bird's eye chili and green spur chili. To reduce the hotness of the spur chili, remove the placenta and all seeds and soak the chili meat in salt water. Chili leaves are used as a source of green color. Cut fresh chili leaves into small pieces, then mash with coconut milk using a mortar and pestle.

Basic curry ingredients: Meat, pork lard, garlic, green chili, holy basil leaves, fish sauce, sugar.

Figure 7.4   (a) Spices in green curry paste. (b) Mashed chili leaves with green chili peppers. (c) Green curry paste. (d) Stir-frying of green curry paste. (e) Green curry.

General curry cooking procedure: Heat pork lard until oil is released in a saucepan. Stir fry meat with garlic until a pleasant aroma is developed, then add the curry paste and season with fish sauce. Add meat stock to desired consistency and add more fish sauce to taste. After simmering, before serving, add green chili, holy basil leaves, kaffir lime leaves. Lastly, add sugar to desired sweetness.

7.3.2 Contemporary Recipe

Green curry paste ingredients:

Herbs; lemongrass, mature galangal, kaffir lime rind, coriander
roots, green spur chili, green bird's eye chili, garlic, shallots,
Spices: coriander seeds, cumin seeds, white peppercorns
Colorant: chili leaves
Fermented seasoning: fermented shrimp paste (*ga-pi*)

Curry ingredients:

Main bites: meat (chicken, pork, beef, fish ball, shrimps, etc.),
vegetable (bamboo shoots, eggplant, Thai eggplant, etc.),
sweet basil leaves
Soup base: coconut cream
Seasoning: fish sauce, palm sugar
Garnish: chili pepper slices, sweet basil leaves

Instructions:

- Maximize spices aroma by roasting coriander and cumin
seeds (separately).
- Cut, slice, or shred herbs ingredients to shorten the plant
fiber, making it easier to pound into a paste
- Blend all curry paste ingredients in an electric blender or use a
mortar and pestle to pound the ingredients from the driest to
the moistest; motion includes pressing and swirling the mortar.
- The chili leaves can be chopped and extracted green color with
diluted coconut milk or pounded with other curry paste ingredients.
- Prepare the curry soup base by heating curry paste in coconut cream.
  - If an oil layer is desired, heat coconut cream a little bit at a
  time to maximize the emulsion breakdown. Once enough
  oil is released, add the curry paste.
  - If time is limited, stir fry curry paste in vegetable oil before
  adding coconut milk.
  - If a thicker, creamier body of the soup is preferred, heat
  coconut cream and the curry paste simultaneously. More
  coconut cream can be added towards the end of cooking.
- Once the curry paste develops a very intense smell, add meat
and vegetables—season with fish sauce and palm sugar to
taste. Garnish with sliced chili pepper and sweet basil leaves.

7.3.3 Culinary Aspects

Today's green curry (*gaeng-keow-wan*, แกงเขียวหวาน) was not much different from the MKHP recipe. There is an explanation in MKHP cookbook that the differences between red curry and green curry are (1) there is no dried chili in green curry, and (2) the color of the chili used is only green. To add a more intense green color, remove the seeds and placenta from chili, soak in brine for a while, then strain and leave for air-drying before using the rest of the chili paste ingredients. The addition of chili leaves also gives an intense green color to the chili paste. However, we found it hard to obtain green chili leaves.

The fresh market where the author went shopping for ingredients is the biggest fresh market in Thailand, the *Or-Tor- Kor Market* (ตลาด อตก). This market has all ingredients for Thai foods that are hard to find in other markets. Because of the size and varieties of fresh ingredients and food stalls, the market was ranked fourth of the world's best fresh market (Goldberg 2017). Nonetheless, chili leaves were not available there as an ingredient. Luckily, the whole plants are available in the nursery that sells plants. So the team bought the entire plant only to use its leaves. The low demand for chili leaves is probably because most people no longer prepare curry paste from scratch and because the curry's brownish color is acceptable.

While pounding for curry paste in the mortar, the color is so brightly green hue with great intensity. However, due to instability of chlorophyll pigments, the green color does not last very long. First, chlorophylls turn the creamy soup green as the pigments are fat-soluble and then turn to dull browning green (chlorophylls turn into pheophytin pigments) with the heat and acidity of the soup. Slices of spur chili pepper and sweet basil leaves can be used for decoration in the serving bowl.

Thai curry varieties share the basic ingredients of culinary herbs and spices in the chili paste. Both red curry (*kaeng-ped* or *kaeng-daeng*) and green curry (*gaeng-keow-wan*) are made with almost identical chili paste. The only difference is the type of chili used. In red curry, dry red chili is used. Traditionally, a chili variety from *Bang Chang* area with larger fruit is used because of its vibrant color. For the green curry, of course, only fresh green chili is used. To enhance green color, a water extract of chili leaves can be added. To prepare curry paste for *pa-naeng* curry, simply add coriander seeds and roasted peanuts to the red curry paste. Human recognizes and identifies flavor as a whole (not by individual ingredients), so most people are not aware of how similar these curry paste ingredients are.

7.3.4 Science Aspects

The first step of making curry is to prepare the curry paste. Coriander seeds and cumin seeds will gain more aromas if roasted, just like roasted coffee bean will have more aroma compounds developing from the Maillard reaction. Use low heat, stir regularly to avoid burning. Stop the roast once the aroma of roasted coriander seeds and cumin seeds is strong enough. Seeds color only slightly darker and it should take only around a minute since only a teaspoon is needed.

If recipe calls for dried chili, it should be soaked in water to hydrate and soften, cut into smaller pieces and seeds to be removed. In the mortar, start with chili since it is fibrous and takes more work to macerate into a puree consistency. Salt crystal, preferably large size, is added with chili to help breaking down the fibrous chili skin is believed to help reduce splash. Then, followed by hard and dry spices, the fresh and delicate herbs are added last. Essential oils gradually release from the blunt force of each pounding. Shallot and garlic should be the last of the pounding process to allow just the right time for enzymatic reaction and allicin to develop.

Modern food blender in place of pestle and mortar can be used though aroma and consistency of the paste can be different. The blunt force is more efficient in breaking the membrane of oil glands and releasing essential oil, as well as other enzymes and substrate that are stored in different organelles within plant cells (as detailed in garlic topic). There can be a physical limitation to using the blender. Because of how little the ingredient is required, it might not reach the blade and got left unmixed. Commercial curry paste is readily available, but as the aromatic essential oils are highly volatile, some aroma components may have been lost during storage.

Since the aromatic essential oils are fat-soluble, lipid or fat content in coconut milk can act as a solvent to facilitate the release of these aromas. As discussed in the topic of coconut milk, coconut milk is an oil-in-water emulsion. Hence, the structure will have to be broken by heat to release pure oil. It is easier to control the amount of oil released when higher fat coconut cream is used. So, use the first squeeze of the grated coconut meat (*hua-ga-ti*) for stir-frying the curry paste. As for the shelf-stable processed coconut milk products, such as canned or UHT, it might take longer to break the emulsion. The process might include homogenization or added chemical emulsifier and stabilizer to stabilize the emulsion (preventing separation of oil and water phases). One might refrain from shaking coconut milk can and gradually pour the content out to see if the product is already separated. If so, the top layer has a higher fat content and can be used to start the curry making. Another method is starting with vegetable oil to let the oil work as a solvent, extracting volatile compounds.

If tough cut such as the shoulder or the one containing a lot of connective tissue (collagen) is used, meat should be simmered with diluted coconut milk in a separate saucepan to ensure the tenderness. Tough collagen will be broken down as the cross-link between amino acids chain is heat-sensitive. The shorter chain becomes solubilized and converted to soft and soluble gelatin with moisture and low heat (50–71°C).

On the other hand, the tender cut of meat, such as chicken breast or tenders, is easily overcooked and dry. The protein in these cuts will get denatured and precipitated, causing the muscle to lose the ability to hold water. The meat shrinks and toughens as the muscle fiber gets shortened. There are two types of muscle protein, myosin precipitates first at 55°C, then when the temperature reaches 70–80°C, actin precipitates—any fat in the muscle melts and releases. Thus, prolonged heating in the tender cut of meat should be avoided (Vaclavik and Christian 2014).

Some vegetables can turn brown from enzymatic browning reactions. Polyphenolic compounds in some vegetables can be oxidized and turn brown by polyphenols enzyme when exposed to oxygen (Lee and deMan 2018). Thus, these vegetables, especially eggplant, will turn brown when cut and exposed to the air. To avoid the undesirable dark color in curry, the pieces of eggplant should be kept submerged in water till ready to cook. As the enzyme can be inactivated with heat, the eggplant should be added to boiling curry. During this step, all surfaces of the eggplant will have to be totally submerged in the hot curry. If some surface area is left exposed to the air above the curry, that area can be oxidized and turn brown before the enzyme is deactivated. Once the enzyme is inactivated, the eggplant will not turn brown even when exposed to oxygen.

With great varieties of herbs and spices used, the curry paste contains an abundance of phytochemicals, which simply refers to chemicals from plants. The phytochemicals are mostly polyphenolic compounds, which have been shown to have high antioxidant activities. Thus, they have led to many potential health benefits, significantly reducing the risks of diseases associated with oxidative damage from free radicals, such as cardiovascular diseases, cancer, type 2 diabetes, and osteoarthritis. Considering the definition of "functional food" as food with additional benefits beyond basic nutrients, herbs and spices might be regarded as functional food. Thai food culture was mentioned using culinary herbs and spices for health benefits (Embuscado 2015; Tapsell et al. 2006). With bioactive compounds identified in more herbs and spices consumed as food in Thailand, their antioxidant, anti-inflammatory, antimicrobial, antihyperglycemic, and cholesterol-lowering effects may contribute to anti-aging benefits. The addition of coconut milk curry also helps extract both water-soluble and fat-soluble bioactive compounds from curry paste, potentially increasing the herbs and spices (Khanthapok and Sukrong 2019).

7.3.5 Modification and Variation

For the red curry, the dried chili has to be soaked in water until it becomes soft. Cut in pieces, discard seeds inside. Pound the chili and crystal salt first, followed by other ingredients just like the green curry.

To add a warmer tone to green curry, add a small amount of turmeric (*ka-min*) to the curry paste.

There are two types of garlic widely available. The locally grown Thai garlic has much smaller cloves but is very pungent and aromatic. Lately, imported garlic from China has become available in the market at a much cheaper price per kilogram (Pisanwanich 2019). Chinese garlic has much bigger cloves but is not as aromatic and spicy, so Thai garlic is especially preferred especially in curry paste.

If a less-tender cut of meat is used, like shoulder or tendon, simmer long and slow to convert tough collagen to gelatin. When cooking with a tender cut of meat like chicken breast or tenderloin, heat just enough to cook.

The curry paste can be stir-fried in a separate pan to avoid excessively long heating of coconut cream and allow the emulsion to break down. Then, transfer the mixture into coconut milk in another saucepan. The trade-off is the lacking of coconut aroma.

If a vegetable that is susceptible to enzymatic browning is used, e.g., eggplants, adding vegetables when the curry is boiling, submerging all the vegetables under soupy curry till cooked.

To make *pa-nang* (พะแนง), add coriander seed and roasted peanuts to the red curry paste. This type of curry has a much creamier and thicker consistency. So the breaking down of coconut emulsion is not as much. The *pa-nang* is a meat-only dish, no vegetable except for the garnish.

To improve the nutritional profile, use less salt and fish sauce to reduce the sodium content. Use more diluted coconut milk to reduce saturated fatty acids from coconut. Keep in mind that curry is just a part of a meal. It is to be eaten with various vegetables as a part of the Thai dining experience. Choose to pair the curry with steamed unpolished rice for extra benefits of fiber and vitamins from rice bran layers and essential fatty acids from rice germs.

Coconut oil has a melting point of 24–25°C. Thus, the oil and the curry sauce will turn solid after refrigeration. Simply reheat, and the curry will get back to liquid.

Curry paste can be kept for a long time, especially in low storage temperature, due to antimicrobial activities in the spices. However, the trade-off is a loss of aroma.

The meats commonly used in green curry are chicken, pork, beef, and processed meat such as fish cake (*loog-chin-pla*, ลูกชิ้นปลา). Sometimes chicken offals and blood cubes are added along with chicken meat while it is less common to have other animal's offals and blood cubes in the soup when other meats than chicken are used.

The modern version of green curry soup is also with Thai eggplants and pea eggplants, or winter melon. Galingales, cut
into strips, are also commonly added into the curry, which
add flavor and some hotness.

7.4 Massaman Curry

Massman curry (*gaeng-mut-sa-mun*, แกงมัสมั่น) is the most complex curry.
In addition to the basic curry paste, it requires additional spice such as
mace, cinnamon, clove, and cardamom. Also, in addition to spiciness,
saltiness, and sweetness, it also requires sourness from tamarind.

7.4.1 Vintage MKHP Recipe (Figure 7.5)

Curry Paste Ingredients: Dried chili, shallot, lemongrass, coriander
roots, fermented shrimp paste, mace, cardamom, cinnamon, pepper,
cumin, coriander seeds.

Figure 7.5   (a) Massaman curry paste ingredients. (b) Massaman curry paste. (c) Bitter orange
(*som-za*, ส้มซ่า). (d) MKHP vintage recipe massaman curry.

Curry ingredients: lard, chicken cuts, coconut milk, fish sauce, sugar, tamarind juice, onion, roasted peanuts, betle leaves (*bai-ploo*, ใบพลู), cinnamon, bitter orange juice, raisins

Instruction:

Curry paste preparation: Roast all ingredients till fragrant, then pound until smooth to a paste-like consistency.

Cooking the curry: Stir-fry chicken cuts with lard until slightly brown; discard excess oil. Add massaman curry paste, fish sauce, and coconut cream.

Add more fish sauce, palm sugar, and tamarind juice to achieve the desired saltiness, sweetness, and sourness.

Add whole grilled peanuts. Simmer till chicken is cooked to desired tenderness, then add onion.

Lastly, before turning off the heat, add bitter-orange juice, raisins, and betle leaves.

7.4.2 Contemporary Recipe (Fig. 7.6)

Common massaman curry paste ingredients:

Herbs: spur chili, salt crystal, shallot, garlic, lemongrass, mature galangal, coriander root,
Spices: white peppercorns, nutmeg, mace, cumin seeds, cloves, cardamom, cinnamon
Fermented seasoning: *ga-pi*

Common massaman ingredients:

Meat: chicken, beef, etc.
Vegetables: potatoes, onion, roasted peanuts,
Soup base: coconut cream
Seasonings: tamarind juice, fish sauce, palm sugar
Garnish: roasted cardamom, bay leaves,

Common instructions:

- Maximize spices aroma by roasting nutmeg, mace, cumin seeds, cloves, cardamom, and cinnamon (separately).
- Create roasted flavor profile of herbs (shallots, garlic, lemongrass, galangal, coriander roots by roasting till aromatic, then finely chopped to save the pounding time. Roast the shrimp paste also.

Figure 7.6   Contemporary massaman curry
Photo credit Ruen Mallika

- Blend all curry paste ingredients in an electric blender or use a mortar and pestle to pound the ingredients from the driest to the moistest; motion includes pressing and swirling the mortar.
- Cook meat in coconut milk, simmer till tender.
- In another saucepan or pan, prepare the curry soup base by heating curry paste in coconut cream.
- Once the curry paste develops a very intense smell, add the stir-fried curry paste to the other saucepan with chicken. Add meat, potatoes, roasted peanuts, roasted nutmeg, and bay leaves, and simmer. Adjust to preferred taste with tamarind juice, fish sauce, and palm sugar to taste.

7.4.3 Culinary Aspects

The re-creation of the MKHP massaman recipe was performed with a traditional *ung-lo* stove, charcoal for heat, and clay pottery for a saucepan. We found that it took a very long time to heat up the clay pottery, and even with the long wait, the temperature could not go very high. As a result, it was not an efficient way to brown the chicken. The recipe does not require oil from coconut milk, so no extra heat is required to break the coconut milk emulsion.

As for the curry paste, the proportion of dried chili is relatively high. Thus, the overall color seems to be intense red compared to the more modern recipe. At the same time, the aroma of roasted spice was less dominant. All herbs and spices should be crushed with the pestle to fine pieces. The fine particles will allow homogenous paste, distributing the flavor and increasing the surface area for flavor release. The curry paste must have a puree consistency; otherwise, the floating bits of herbs and spices can be considered embarrassing.

There were a lot of roasted peanuts required, so the finished curry had a distinctive texture and flavor of peanuts. Another difference from todays' massaman curry is the addition of raisins. The sweetness and sourness of raisin bear a resemblance to the roasted duck curry (*gaeng-ped-pade-yang*, แกงเผ็ดเป็ดย่าง). Because of herbs and spices, the earthy, smoky aroma of clay and charcoal was not detectable. This is why the clay pottery nowadays is used to cook or serve rice only as the aroma can be detected from the mild-aroma-cooked rice.

The meat in massaman is almost always chicken or beef. The vegetables can be several types of yams, potatoes, or any tubers with a firm texture. The contemporary version of massaman curry nowadays do not contain raisins.

7.4.4 Science Aspects

All the recipes in the top tier of the Gub-Kao Grid (boxes 1, 2, 5, and 6 in Fig. 6.3) require the pounding of herbs and spices into curry paste which will become the flavor foundation of each dish. The massaman curry is on the top right corner, meaning it has the highest flavor complexity of herbs and spices and is a soupy dish with high-fat content. Roasting spices are required to maximize the aroma potential of the massaman curry.

The meat has to be fried to obtain the browning and aroma, too. This type of curry has much higher fat than another coconut-based curry. Besides massaman chili paste and meat, the other main ingredients are onion and starchy vegetables such as potatoes or yam. This curry is thus very high in fat, and the fatty taste and mouthfeel can be overwhelmed. Thus, a popular accompanied dish to the massaman curry is cucumber salad called *ar-jard*. The sweet, tangy taste with crunchy and firm texture of the salad helps refresh and clean the palates, thus increasing the pleasure of enjoying the fatty curry.

The dish has a heavy texture with very thick curry sauce. The taste is salty, sweet from palm sugar, and also tangy from tamarind water. To prepare tamarind water, simply soak the tamarind in water till the water becomes very dark and thick, then discard the tamarind.

7.4.5 Modification and Variation

The meat in massaman is almost always chicken or beef. In MKHP, there is a suggestion to replace coconut cream with butter and ghee. The massaman with these dairy ingredients would resemble Indian-style curry. Another curry quite similar to massman curry is called **gaeng-ga-ri** (แกงกะหรี่). Its curry paste has fewer spices but contains curry powder (similar to Indian curry powder). The common solid ingredients are meat, onion, and potato, but the *gaeng-ga-ri* does not require roasted peanuts and tamarind juice.

7.5 Som-tum (Spicy Green Papaya Salad)

The first time som-tum recipe appeared is in the *Tumrub-Sai-Yaowapa* cookbook. The dish was to be eaten with coconut-milk rice (*khao-mun-som-tum*, ข้าวมันส้มตำ). There was a suggestion for side dish of this menu, including raw vegetables, crispy deep-fried vegetables and edible flowers, sweetened fried beef floss (*nua-foy-pud-wan*, เนื้อฝอยผัดหวาน), fried fish, chicken curry, or beef curry.

7.5.1 TSY Recipe

Main ingredients: green papaya, garlic, pepper, dried shrimp.

Seasoning ingredients: tamarind juice, palm sugar, fish sauce, diced lime fruit, bird's eye chili is optional

Instruction: Peel off the papaya skin, and wash well to remove any sticky resin. Shred the papaya and mix with salt; remove any excess water.

Pound garlic and peppercorn till smooth.

Pound dried shrimp till the texture is soft and light, then mix with the shredded papaya.

Mix all ingredients well and season with tamarind juice and diced lime fruit for sourness. Balance the sourness with the saltiness from fish sauce and sweetness from palm sugar as desired.

If spiciness is preferred, add crushed bird's eye chili.

7.5.2 Contemporary Recipe (Figure 7.7)

Ingredients: Shredded green papaya; dried or fresh chili pepper; garlic; nam pla-ra; palm sugar; lime juice.

Figure 7.7   Example of contemporary *som-tum*

Photo credit to Ruen Mallika

Optional ingredients: Fish sauce, pickled crab, various types of Thai eggplant (*ma-kua-proe*), tomatoes, yardlong beans (*tua-fug-yao*), peanuts.

Instructions: Pound garlic and chili in pestle and mortar. Add *nam-pla-ra*, *nam-pla*, palm sugar, and lime juice in the mortar and gently mix well with the garlic and chili using pestle and spoon.

7.5.3 Culinary Aspects

This regional dish is a northeastern-style green papaya salad. The distinct flavor comes from *pla-ra*. *Pla-ra* can come in the form of whole fermented fish. *Nam* means liquid or water in Thai language. Therefore, when the recipe is called for *nam-pla-ra*, it is an indication that only the liquid part will be used. Because of strong *pla-ra* aroma, both *pla-ra* and *nam-pla* types of fish sauce may be used in the *som-tum-pla-ra* recipe. If *pla-ra* is totally replaced with *nam-pla* or used in addition to the fish sauce, the recipe is simply called by a generic name *som-tum-pla-ra*. Other essential ingredients include dried shrimp, roasted peanuts, and yardlong beans.

7.5.4 Science Aspects

Umami taste will arise from a combination of the fish sauce and other protein ingredients such as tomatoes and pickled crab or dried shrimps. However, the MSG food additive, known locally as *pong-choo-ros* (literal translation is "flavor-enhancer powder"), is usually added for extra umami and a well-rounded taste. Both *pla-ra* and *nam-pla* are potential sources of umami and kokumi taste, giving the dish longer lasting deliciousness (Kuroda and Miyamura 2019; Phewpan et al. 2019).

The frequency or hotness of chili pepper can be adjusted by the type and quantity of chili peppers and the force used during pounding. The more vigorous the pounding, the higher the amount of capsaicin coming out, and the higher astringency will be distributed evenly.

7.5.5 Modification and Variation

Slightly yellowish papaya can be used in place of green papaya, especially in the rural area where vitamin A deficiency is still an issue.

The development of carotenoids in the papaya gradually changes color from pale green to yellow and orange. These carotenoids, including β-carotene, β-cryptoxanthin, and lycopene, have high bioavailability. Some carotenoids, especially carotene, are provitamin A as they can be converted to vitamin A (Schweiggert et al. 2014). Papaya texture may be softened depending on the degree of ripeness. The softening is caused by degradation and increase in solubility of polysaccharides, especially pectin, on the fruit cell wall (Manrique and Lajolo 2004; Shiga et al. 2009). The softer texture may be right for the elderly as it is easier to chew, while most people prefer the firm texture of the shredded papaya. The shredded papaya can also be replaced by other firm-texture vegetable and fruits to create a variety of this spicy salad, including apple, carrots, santol, cucumber, and sunflower seed sprouts.

The *som-tum* is always served with a variety of fresh vegetables. Sticky rice is a more popular accompanying dish than coconut-milk rice. Since there is almost no-fat source on the menu, carotenoids (provitamin A) absorption can be problematic. In order to improve the nutrient intake of a meal, the *som-tum* should also be accompanied by other meat products. Grilled chicken or *gai-yang* is among the most popular accompanied dish.

7.6 Shrimp Paste Chili Dip

Chili dip or *nam-prig* (literal translation is chili water, น้ำพริก) is another *gub-khao* that is honored as a national heritage. As the name implies, it is concentrated and spicy and always have to be eaten with vegetables. Currently, the most popular chili dip is *nam-prig-ga-pi* (น้ำพริกกะปิ). There are many predecessor versions of chili dip made from fermented shrimp paste (*ga-pi*, กะปิ) before it became *nam-prig-ga-pi* as we know today. On MKHP, the closest recipe is *ga-pi-prig* (กะปิพริก). Then, on TSY, there is *nam-prig* with similar ingredients but it is called *nam-prig-ma-kua-puang* (น้ำพริกมะเขือพวง) because of the addition of the pea eggplant (*ma-kua-puang*). The exact term of *nam-prig-ga-pi* is found in RD1982.

*7.6.1 Vintage MKHP Recipe for **Ga-pi-prig** (กะปิพริก) (Figures 7.8 a–d)*

Ingredients: fermented shrimp paste, spur chili, bird's eye chili, seasonal sour fruits such as hairy eggplant (*ma-ueg*, มะอึก), sour green

Figure 7.8 (a) *Ga-pi-prig* ingredients. (b) Sour fruits ingredients (clockwise, *ma-ug*, *ma-dun*, *ta-ling-pling*). (c) Dried freshwater fish (*pla-sa-lard*) (d) *Ga-pi-prig* chili dip.

mango, plum mango (*ma-prang*, มะปราง), sour garcinia (*ma-dun*, มะดัน), palm sugar, dried shrimp or dried fish.

Procedure: Mash dried shrimp and salt, followed with garlic, fermented shrimp paste, palm sugar, spur chili, freshly squeezed lime juice, and other sour fruit. Garnish with sliced chili pepper. The consistency should be very thick. The dip is customarily served with vegetables, sweetened pork stew, hard-boiled eggs, or grilled catfish.

7.6.2 RD1982 Recipe for Nam-prig-ga-pi

Ingredients for thin (runny) nam-prig-ga-pi: Fermented shrimp paste (*ga-pi*), garlic, red-, green-, and yellow spur chili (*prig-chee-fah*, พริกขี้ฟ้า), bird's eye chili (*prig-kee-noo*, พริกขี้หนู); fish sauce, lime juice, sugar.

Ingredients for (viscous) nam-prig-ga-pi: Same as the thin type with the addition of ground dried shrimps or grilled sun-dried fish, hairy eggplant (*ma-ueg*, มะอึก), eggplant.

Instructions: For thin *nam-prig-ga-pi*, cut spur chili into small pieces (3–4 pieces per chili). Pound garlic and *ga-pi* in mortar with pestle. Add bird's eye chili, then fish sauce, lime juice, and sugar. Mix well. For thick type, add the additional ingredients after adding bird's eye chili. The thin type is accompanied by steamed vegetables while the thick type goes well with raw vegetables.

7.6.3 Culinary Aspects

The chili dip can be made thick, or slightly runny consistency depends on personal preferences. The dip is typically served as a set with omelet, fried fish, a set of blanched vegetables, and fresh vegetables such as morning glory, Thai eggplant, cucumber, and edible flowers. There are wide varieties of fermented shrimp paste, which depend on raw materials, bacterial culture, and method. Each person has their own preference for the shrimp paste (color, flavor, texture, etc.). Unlike fish sauce, there is no recognized brand of shrimp paste. The productions are mostly small scale and likely referred to by the district it came from, such as *ga-pi* from the *Klong-Kone* district of Samut Songkhram province.

7.6.4 Science Aspects

Unlike fermented freshwater fish (*pla-ra*) and fermented saltwater fish sauce (*nam-pla*), *ga-pi* has low water content (semi-solid consistency). The process required draining of liquid from the krills (plantonic shrimps, much smaller than shrimps) and sun-drying. Flavor developed during the fermentation from natural bacteria similar to fish sauce. Many similar chili-dip recipes require low-heat grilling of the *ga-pi*, which is wrapped inside banana leaves. The dry heat generates another set of volatile compounds, likely to be from the Maillard reaction.

Traditionally, the accompanying assorted vegetables can be fried, fresh, or steamed. The steamed vegetable can be topped with coconut cream or prepared by boiling in diluted coconut milk. The addition of fat sources in the meal may help clean the palate by washing the fat-soluble capsaicin from the receptors on the tongue.

7.6.5 Variation and CNW Nam-prig-long-rua

This type of chili dip is now called nam-pig-ga-pi, and the common ingredient for sourness is almost always lime juice. The addition of other raw materials is regarded as a different dish, e.g., mango chili dip (*nam-prig-ma-muang*, น้ำพริกมะม่วง) when replacing lime juice with shredded sour green mango, shrimp chili dip (*nam-prig-koong-sod*, น้ำพริกกุ้งสด) when diced cooked shrimp are added.

In CNW, the nam-prig-ga-pi can be transformed into a more sophisticated version called *nam-prig-long-rua*. Essential ingredients besides the nam-prig is sweetened pork (*moo-wan*, หมูหวาน). The crucial accompanied dish is crispy snakehead fish flakes and assorted vegetables. To cook, fry garlic until aromatize, add *nam-prig-ga-pi*, and stir-fry until cooked. To serve, line a bowl with sweetened pork, crispy fish flakes, sliced pickled garlic, and salted egg yok bits, then mix with the stir-fried *nam-prig-ga-pi*.

The fermented shrimp chili dip can be transformed into a type of curry called **gaeng-run-juan** (แกงรัญจวน), which was first described in the CNW book. The first step is to simmer the meat with lemongrass, shallot, and garlic. Once the meat is tender, add the *nam-prig-ga-pi*. When ready to serve, add freshly squeezed lime juice, lemon grass slices, and sweet basil leaves.

7.7 Stir-Fried Cabbage and Shrimp (*Ka-lum-plee-pud-koong*, กะหล่ำปลีผัดกุ้ง)

7.7.1 MKHP Stir-Fried Cabbage and Shrimp (Figure 7.9)

Ingredients: chopped cabbage, shrimps, sliced pork belly, garlic, fish sauce, sugar.

Instruction: Heat sliced pork belly till lard melts to oil. Add garlic to the hot oil till slightly brown, then add shrimps. Before shrimps are fully cooked, add cabbage. Season with sugar and fish sauce to desired taste.

7.7.2 TGKSLL Stir-Fried Pak-choy

Main ingredients: Vegetable (e.g. pak-choy), meat (e.g. shrimp, chicken), garlic, pork lard

Seasoning ingredients: fish sauce, soy sauce, sugar

Functional ingredients: tapioca flour to thicken the sauce, ashes (usually the left-over from burning logwood at the stove, containing

Figure 7.9   MKHP stir-fried cabbage with lard and shrimps

compounds such as potassium and calcium hydroxide, which give alkalinity to water solution)

Procedure: Cut and prepare the vegetable as desired. The vegetable pieces are to be blanched in alkaline water obtained by dipping a cloth bag of ash into the boiling water. The blanched vegetables should be strained well to remove excess water, which may cause the stir-fry to be too watery.

Prepare the meat ingredients according to preference (size, cuts, etc.) Stir-fry the garlic, meat, and vegetables, and season to taste with fish sauce, soy sauce, and sugar. The mixture of tapioca flour in water is added last to thicken the liquid.

7.7.3 Culinary and Science Aspect

The re-created MKHP stir-fried cabbage with shrimps looked and tasted just like the modern stir-fried vegetable. The use of lard instead of the modern processed vegetable oil gives slightly meaty aroma to the dish, which was harmonized with shrimps and fish sauce very well.

There are basically two types of stir-fried vegetables: A traditional one without thick sauce like in the MKHP recipe and the other one is with the addition of starch solution, which will gelatinize with heat giving the thick consistency to the sauce. Modern version of the stir-fried vegetable dishes may include oyster sauce, which will give thickness and flavor to the sauce. The higher viscosity helps the sauce to stick to the vegetable pieces better.

Depending on the texture of the vegetable used and the diner's preference, vegetables can be blanched to ensure homogenous doneness of the vegetable. The addition of alkalinity to the blanching water, like in the TGKSLL recipe, helps turn chlorophyll pigments to chlorophyllin pigments that are bright green. However, the higher pH may dissolve the cellulose structure of the vegetable, leaving it undesirably mushy. Thus, this method works well with very thick and firm vegetables like the bok choy stem.

Seasoning with fish sauce imparts deliciousness from umami taste and fish sauce aroma. The cooked fish sauce yields the most acceptable aroma when cooked in high heat. Thus, the fish sauce should be added toward the last step and when the food is very hot. Another technique is to sprinkle the fish sauce directly onto the side of the metal wok. This technique will give a slightly burnt aroma. The differences in the scent of these techniques may involve the peptides, degree of the Maillard reaction, and possible pyrolysis.

Recently, the oyster sauce has been a popular seasoning for stir-fry dishes, both for vegetables and meat. For best results, oyster sauce is the last ingredient to add. It gives a savory, meaty aroma, dark brown color, and starch-thicken sauce to the finished product.

7.8 Gaeng-liang

Gaeng-liang has a soup consistency but differs from regular clear soup in that it requires curry paste (kruang-gaeng) as a flavor foundation.

7.8.1 MKHP Gaeng-liang (Fig. 7.10)

Ingredients in curry paste: snakehead fish (pla-chon, ปลาช่อน), black peppercorn, shallot, galingales, fermented shrimp paste (optional).

Ingredients in the soup: banana blossom (hua-plee, หัวปลี), ridge gourd (buab, บวบ), ivy gourd leaves (tum-leung, ตำลึง), star gooseberry leaves (ma-yom, มะยม), young pumpkin leaves and pumpkin flowers,

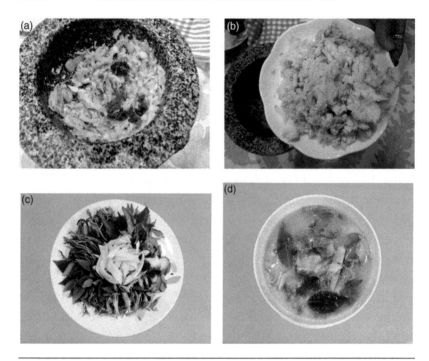

Figure 7.10   (a) Pounding of *gaeng-liang* curry paste. (b) Cooked freshwater fished (*pla-chon*), minced. (c) Various vegetables. (d) *Gaeng-liang*

Opiliaceae plant leaves (*pug-wan*, ผักหวาน), lemon basils (*bai-mang-lug*, ใบแมงลัก), stock.

Procedure: The MKHP cookbook described only the curry paste for *gaeng-liang*, but no detailed instructions. In general, the first step is to pound all herb and spice ingredients into a paste-like consistency (curry paste). Next, the snakehead fish cuts are boiled in a separate saucepan. The cooked fish is minced by hand or with a fork, then transfer the minced cooked fish to the mortar to mix well with the curry paste. To cook, boil the stock, then add the curry paste mixture. Bring to boil, add vegetables in the order of heat requirement, and season with fish sauce.

7.8.2 Modification and Variation

Gaeng-liang is milder than other curries because the spiciness comes only from pepper, no chili or dried chili, and no fatty ingredients like coconut milk. Instead, the curry paste mixed with minced fish is the flavor foundation for this soup. The pleasantness comes from refreshing herbs combined with umami from minced fish. This *gub-khao* is especially versatile because its flavor profile can accommodate any combination of vegetables, and the meat can be any freshwater fish or shrimp. Typically, there is a firm texture like pumpkin and banana blossom; a soft and spongy one like ridge gourd, snake gourd, young watermelon, or mushroom; and leafy vegetables like ivy gourd leaves and any seasonal indigenous plant leaves. The aroma from galingale and fermented shrimp paste is optional, while the aroma of lemon basil is a must, or the soup cannot be regarded as *gaeng-liang*.

Gaeng-liang is a type of soupy dish with very low fat and high chemesthesis, thus can be placed in Box 5 of the Gub-khao Grid (Figure 6.3). Other soups in this category are, for example, *gaeng-som* (แกงส้ม). The curry paste does not require heating with oil or coconut milk, so it can be described as fresh curry paste (เครื่องแกงสด). *Gaeng-som* curry paste is very similar to *gaeng-liang*, just with the addition of dried red chili. Chili turns the soup color to orange and gives a higher degree of spiciness. This soup contains various vegetables similar to *gaeng-liang*, but with the addition of tamarind juice for a very sharp sourness. The word "*som*" signifies the color of the soup (orange) and the sour taste.

7.9 Red Curry

7.9.1 *MKHP Chicken Curry (Gaeng-ped-gai,* แกงเผ็ดไก่*) (Fig. 7.11)*

Curry paste ingredients: dried spur chili, lemon grass, salt, fermented shrimp paste, coriander roots, kaffir lime zest, coriander seeds, cumin, red onion, garlic, galangal.

Chicken stock ingredients: chicken bones, shallot or onion, radish (daikon), peppercorns, cloves, salt

Curry solid ingredients: chicken meat and cuts, chicken offals (kidney, liver, stomach, heart).

Procedure:

Curry paste preparation: Pound all ingredients until achieving a smooth paste-like consistency.

- Chicken stock preparation: Boil chicken bones in water. Add shallot or onion, radish, peppercorns, and cloves to increase savory richness if desired—season with salt.

Figure 7.11   (a) Red curry paste ingredients. (b) Dried red chili (soaked, placenta removed, cut). (c) Red curry (*gaeng-ped-gai*).

- Cooking curry: Heat pork lard in a saucepan till hot, and add crushed garlic. Once the garlic turns light brown, add curry paste and stir, then add the meat and offals.
- Season with fish sauce, adjust to desired consistency with chicken stock and add green chili, sweet basil, kaffir lime leaves, and sugar as desired.

Note: **Red curry** (*gaeng-ped*, แกงเผ็ด) refers to a type of soupy gub-khao that requires curry paste, and the paste contains both herbs and spices (others, such as gaeng-liang, have only herbs). *Gaeng-ped* requires curry paste and fat sources (coconut milk, lard, or other oils); thus, they are high in both chemesthesis and fat. This *gub-khao* has a high complexity of ingredients and flavor profile and fits in box 7 of the *Gub-khao* Grid (Fig. 6.3) along with other curries such as green curry and massaman curry. As mentioned in topic 5.7, the curry cooking technique, the curry paste needs to be very fine and smooth. The curry paste is also required to cook with fatty ingredients (either oil or coconut milk) until the aroma is pleasant and intense before adding other solid elements.

7.9.2 *MKHP Chicken Curry with Coconut Cream*
(*Gaeng-gai-gub-ga-ti*, แกงไก่กับกะทิ) *(Fig. 7.12)*

Curry paste ingredients: Same as the *gaeng-ped-gai*, but replacing lard and chicken stock with coconut milk.

Curry paste preparation: Pound all ingredients until achieving a smooth paste-like consistency.

Cooking curry: Heat coconut milk in a saucepan till oil separates, then add curry paste and stir. Once the mixture becomes aromatic, add meat ingredients. Season with fish sauce, adjust to desired consistency with chicken stock, and add green chili, sweet basil, kaffir lime leaves, and sugar as desired.

7.9.3 *Modification and Variation*

As mentioned in Section 7.3.5, modification and variation of green curry, using only red chili, will be the basis of the *gaeng-ped*, which has a distinctive red color. These two MKHP recipes shared the same curry paste. The only difference is the source of

Figure 7.12   Red curry with coconut cream (*gaeng-gai-gub-ga-ti*)

fat (lard or coconut cream). When tested, the re-creation of both recipes, the red curry with coconut cream (*gaeng-gai-gub-ga-ti*, แกงไก่กับกะทิ), was more delicious. The coconut cream harmonized all the ingredients. There was no particular taste or aroma that dominated the curry. Overall, red curry with coconut cream was more delightful to eat, truly well-rounded, and delicious. The chicken curry without coconut cream (*gaeng-ped-gai*, แกงเผ็ดไก่) had more sharp and pungent flavor. The dish seems like an improvised version of the one with coconut cream. However, in modern Thai cuisine, the name *gaeng-ped* (spicy curry, แกงเผ็ด) has become a generic term for many types of curry soup either with or without coconut milk. So, this is a contrast between vintage and contemporary cooking.

The recipes called for one whole chicken, including offals. The use of chicken offals is not usually seen in the current curry. At the time of this cookbook, there was no electricity in Thailand yet. Thus, it may have been more efficient to cook all edible parts right away in the presence of herbs and spices. In addition, the phytochemicals in curry paste give antimicrobial properties to the meat product.

7.10 Spicy Mixed Salad (*Yum-yai*)

The cooking of the *yum* is a simple tossing of any plant-based and animal-based ingredients and seasoned to the balance of saltiness, sourness, sweetness, and spiciness. Thus, there is an endless possibility of creating *yum* dishes. The dish is so versatile that another meaning of the word *yum* in Thai language is medley, mix-and-match, or combination.

7.10.1 MKHP Recipe for Yum-yai (ยำใหญ่) (Fig. 7.13)

Solid ingredients: Any combination of meat, eggs, and vegetables.

Meat-based solid ingredients: Diced pork rind, chopped pork meat, diced lard, offals (stomach, heart, liver), pulled cooked-chicken meat, cooked shrimps, steamed squid, dried shrimps.

Figure 7.13 *Yum-yai*

Dairy-based ingredients:- Hard-boiled duck eggs, sliced duck-egg omelet.

Plant-based solid ingredients: Chopped pickled garlic cloves, thinly sliced cucumber meat (no seeds), thinly sliced radish, cooked bean sprouts, beancurd skin, tofu, chili, bean noodle, black ear mushroom, coriander leaves

Yum dressing ingredients: soy sauce, vinegar, sugar, fish sauce, lime juice, coriander roots, garlic, pepper.

Instructions:

Solid Ingredients: Cooked each ingredient separately to the preferred doneness and texture. Cut or slice into small pieces. Put all ingredients in the preferred proportion in a large mixing bowl.

Yum dressing: Pound coriander leaves, garlic, and peppers until smooth. Season to taste with vinegar, fish sauce, sugar, and soy sauce.

When ready to serve, toss the salad ingredients with the yum dressing until mixed well.

This MKHP recipe of *yum-yai* is not very spicy as the hotness only comes from the pepper. However, the typical *yum* as a *gub-khao* nowadays is expected to be very hot and spicy; thus, it can be called spicy mixed salad. As the dish has no or very low-fat content, this dish fits in Box 3 of the Gub-khao Grid (Fig. 6.3). Another example of yum varieties is *laab-moo*.

Laab-moo (Spicy minced pork salad):
Solid ingredients: Minced pork, sliced shallots, spring onion, saw-tooth coriander leaves
Aromatics ingredients: ground roasted rice, mint leaves
Seasoning: fish sauce, lime juice, ground dried chili
To serve: In a bowl, toss all ingredients, season to taste as desired, and mix well. Serve with a combination of fresh vegetables.

7.10.2 Modification and Verification

For scientific principles of the mixed salad or yum, please refer to topic 5.4, salad. Overall, the re-creation of this dish is another surprise. Our team was impressed by how labor-intensive this dish was. At the time this recipe was written, there was no electricity, refrigerator, or even ice. We came to understand why the dish was considered royal cuisine at the time. There are many meat-based ingredients, like pork, chicken, shrimp, and squid. Each element would require a long time to obtain fresh from the source and manpower to prepare in the kitchen. Just like what we noticed from the chicken curry, this dish calls for offals or internal organs of pigs. To us, this reflects an old day's attempt to be more efficient with animal products, especially the larger land animals.

The mixture of coriander roots, garlic, and pepper used in this MKHP *yum-yai* recipe dressing is a basis for many dishes. They are also known as 3-friends (*sarm-gler*, สามเกลอ) in Thai. Generally, this 3-friends mixture is used with any meat, especially minced meat, to add depth of flavors and mask any off-flavor of the meat. A modernized version of spicy salad dressing tends to replace pepper with fresh chili. For example, in the **yum-nua-yang** (spicy beef salad, ยำเนื้อย่าง), the dressing preparation starts with a pounding of garlic, coriander root, and red spur chili. To add complexity to the *yum*, more contrast of flavor and texture of ingredients can be used. The dressing can be combined with coconut curry and roasted chili paste.

The *yum* dressing can be either transparent and watery, consisting of lime juice, fish sauce, and sugar, or an emulsion-based with fatty ingredients such as coconut cream and roasted chili paste (*nam-prig-pao*). The main difference between Thai-style *yum* dressing and western-style salad dressing is the taste. The *yum* has distinctively sharp sour, spicy, salty, and sweet. The source of sourness can be various kinds of fruits, especially shredded sour green mango. Ingredients such as tamarind juice and palm sugar have darker color and flavor profiles that go well with coconut cream-based *yum* dressing. Examples are banana blossom *yum* (*yum-hua-plee*, ยำหัวปลี) and pomelo *yum* (*yum-som-o*, ยำส้มโอ). Other ingredients that add layers of flavor complexity include lemon grass, shallot, shredded kaffir lime leaves, roasted grated coconut meat, dried shrimps, and roasted peanuts.

7.11 *Pla-ra* Chili Dip

There are many varieties of fermented freshwater fish or *pla-ra*. The differences are in the type of fish, other additional ingredients such as roasted rice, bacterial culture, rice bran, and fermentation duration. Archaeology evidence showed the pre-historical consumption of fermented fish in the area of Thailand. The advanced technology identified several aromatic and taste enhancer compounds in the *pla-ra*. This ingredient continues to be crucial to many Thai cuisine dishes, and the *pla-ra* has been honored as national heritage.

7.11.1 *MKHP Recipe for* **Nam-prig-pla-ra** *(น้ำพริกปลาร้า) (Fig. 7.14)*

Ingredients: dried shrimp, dried chili, garlic, shallot, sliced galangal, *pla-ra*, whole kaffir lime fruit, sugar, fish sauce.
   Instructions:

- Strain *pla-ra* to obtain the fish and discard the liquid.
- In a saucepan, boil the fermented fish with galangal slices till both meat and bones dissolve into the liquid.
- Grill chili and garlic till aromatic.
- In a mortar, pound dried shrimp (or dried fish) with grilled chili and garlic. Add the cooked *pla-ra* to the desired consistency and season with palm sugar to taste.

Figure 7.14 (a) *Nam-prig-pla-ra* ingredients. (b) Boiling of *pla-ra*. (c) Ready to serve *nam-prig-pla-ra*.

7.11.2 Modification and Variation

The MKHP *nam-prig-pla-ra* (น้ำพริกปลาร้า) recipe is not very difficult to follow. All ingredients are readily available in any fresh market. At the time of purchase, the *pla-ra* has a very intense fermented aroma. Some people even describe it as the smell of rotten fish. The whole fish body is clearly visible in the thick fermented juice. The fundamental transformation occurred during the boiling. After just a few minutes, the fish body was almost completely dissolved into a liquid. The disappearance of fish flesh resulted from an enzymatic reaction from bacteria during fermentation (see also Section 4.4). The only solid mass visible is the large fishbone pieces that were later be strained out.

During boiling, the smell was very bold initially, almost unbearable for a someone who experienced this for the first time. Then, upon the addition of galangal, the smell has been softened into a mild and subtle fermented fish aroma. The addition of grilled (burnt) garlic and chili pushes the fermented aroma to the middle note, so slightly charred herbs and aromatic spices are more dominant. The addition of dried shrimp or dried fish imparted body and thickness to the chili dip. Finally, the skin of kaffir lime fruit added a refreshing citrusy endnote to the chili dip, an optional but invigorating to have ingredient. Ultimately, with the last step of seasoning with fish sauce, chili, and palm sugar, the unattractively smelly fish had utterly transformed into the enjoyable *nam-prig-pla-ra*.

This chili dip can be served with various fresh and steamed vegetables, hard-boiled egg and fried fish. Either cooked or fresh, the vegetables are fat-free. As a result, the high concentration of chemesthesis in the *nam-prig*, especially capsaicin from chili, may be too overwhelming. Therefore, a coconut milk-added version is a good alternative. The cooking technique of *nam-prig* and coconut cream is called *lhon* (หลน), and it can be applied to *pla-ra* also. To make *lhon-pla-ra*, boil the *pla-ra* in diluted coconut milk until it dissolves into a liquid form. Then, follow all the steps of *nam-prig-pla-ra* method steps. Finally, add coconut cream and turn off the heat after seasoning to the desired saltiness and spiciness. The coconut cream is a good source of fat in this *lhon-pla-ra* recipe.

The *pla-ra* chili dip and its variations may not be as widely consumed as the *nam-prig-ga-pi* because it usually takes several exposures before one can appreciate the true spirit of this old-time favorite recipe. Nonetheless, the dish is a representation of a genuine and authentic traditional Thai dish.

7.12 Special Note on Pad Thai

The word "pad thai" has just been added to the Oxford dictionary and is described as *a dish from Thailand made with a type of noodles made from rice, spices, egg, vegetables and sometimes meat or seafood*. Though it is recognized worldwide as a representative of Thai food, there was no pad thai on any of the vintage recipes cited in this chapter because this menu item woas founded later around 1940s.

The dish that bears the closest resemblance to the pad thai in the authors' opinion is the crispy rice noodle (*mee-krob*, หมี่กรอบ) (Fig. 7.15a). M.L. Nuang Ninrat mentioned the *mee-krob* dish from her childhood in CHW. Hence, the dish must have existed since the early 1900s or earlier. The *mee* type noodle is a small white round (angel hair) rice noodle. The word *krob* means crispy. Thus, the dish requires deep-frying of the noodles and other ingredients, such as shredded yellow tofu and small shrimp, to the desirable hard-crack crispiness. The

Figure 7.15   (a) *Mee-krob.* (b) Pad thai.

original recipe recorded by Ninrat included fermented soybean paste (*tao-cheow*, เต้าเจี้ยว) and *som-za* (small citrous fruit, ส้มซ่า), but they are not found in most modern *mee-krob*. In addition, just like pad thai, (Fig. 7.15b) the *mee-krob* consists of chives (กุ้ยช่าย), shrimps, eggs, shallot, garlic, red spur chili, coriander leaves, cane sugar, palm sugar, vinegar, and lime juice (Ninrat 1996). Other meat, such as pork and chicken, can be shredded and deep-fried as optional ingredients.

There are two main differences between mee-krob and pad thai. First, *mee-krob* needs crispy noodles, crispy tofu, and small crunchy shrimps, while pad thai is made from thin flat white rice noodles, tufu, and shrimps in their regular cooked texture (with a bit of firmness like al dente pasta for rice noodle). Also, pad thai has additional ingredients not found in *mee-krob*, i.e., sweeten dried Chinese radish (หัวไชโป๊หวาน), sour tamarind paste, and fish sauce. Thus, the taste profile of the two dishes is similar, i.e., sour, salty, and sweet. The difference is in the texture that is crunchy and crispy for *mee-krob*, and flavor that is more toasty in *mee-krob* due to the Maillard reaction from deep-frying of ingredients.

Suggested pad thai recipe, courtesy of Ruen Mullika, the Michelin-star Thai restaurant is shown in Appendix 2. More details on the history of pad thai were mentioned in Chapter 2.

7.13 Conclusion

The re-creation of 11 dishes from vintage recipe has given us an insight into the early twentieth-century kitchen. All savory dishes (*gub-khao*) were designed to be eaten with rice. Interestingly, every recipe contains at least one fermented seasoning. The most popular one is *nam-pla* (fish sauce), found in every recipe in this chapter. Next is *ga-pi* (fermented shrimp paste), though not always noticeable as sometimes it is used only in a small amount as an ingredient of curry paste. Lastly, *pla-ra* (fermented freshwater fish) requires herbs, spices, and the proper cooking technique to transform the smelly fermented fish into a delicacy.

Although chili has become a symbol of Thai cuisine in the eye of foreigners, it is actually not as ubiquitous as fish sauce. The hypothesis is supported by the fact that many of the vintage recipes regarded chili as an optional ingredient. Thus, the non-spicy version of Thai food was well accepted. On the other hand, the salty fermented seasoning cannot be omitted, or the dish's identity will be lost.

The understated dishes that are so common to Thai people and well recognized worldwide, like pad thai and *tom-yum-koong*, did not exist in any early twentieth-century cookbooks. Instead, massaman curry, the origin of which dates back to the sixteenth century, was mentioned with a much different appearance and flavor. The predecessors to today's pad thai, *tom-yum-koong*, and massaman curry seem to have milder flavors. For example, today's *tom-yum-koong* is much more sour and spicy than the *gaeng-nog-mhor*. Thus, these vintage recipes tend to emphasize the milder but well-balanced taste. Those modernized dishes are well balanced too, but each flavor attribute is much bolder.

The time-consuming dish preparation also made us realize that fish and marine animals were more convenient for the tropical lifestyle of Thai people. Better transportation, electricity, and a cold supply chain may be the important turning points that shift Thai cuisine to include more land animals.

The *gub-khao* recipes discussed in this chapter elaborate on key messages revealed in previous chapters. Traced back from prehistoric times, significant components in Thai people's diet have always revolved around rice, fish, and salt. Rice is regarded highly in every meal. Consequently, all other dishes are perceived as less significant and are straightforwardly referred to as "rice accompaniment" or *gub-khao*. Fermentation of fish and salt yielded the umami-rich seasoning ingredient needed in almost every *gub-khao* recipe. Then, influenced by many food cultures and imported ingredients, driven by the open-mindedness of the locals, what used to be innovative dishes of the time have evolved and intertwined to become Thai cuisine identity. The non-native plant such as chili gradually has become very crucial, so much so that its fiery sensation might hold an answer as to

why Thai food has been so popular. Other distinguishable characteristics of Thai food mentioned in earlier chapters, e.g., having a very
high number of ingredients, the use of galangal and lime juice, and
the harmony of sensory stimulation, are also shown through these
recipes. The sensory attributes of colors, flavors, trigeminal sensations, and mouthfeels are appealing enough for any Thai *gub-khao* to
be gastronomically pleasant. Yet, building a meal according to *sum-
rub* principles gives another dimension of flavor balance. In addition,
the traditional style of seating and serving also creates an even more
memorable experience for anyone enjoying a Thai meal.

The science behind the arts of cooking, emphasizing the chemical properties of ingredients, their interactions, and transformation,
offers some plausible explanations for the development of the sensory
attributes unique to Thai national cuisine.

Acknowledgment

The authors appreciated Mr. Chaiwat Phulsawat and Mr. Worasak
Yangon for contributing to the analysis and recreation of the MKHP
recipes. Also, sincere gratitude to Ms. Chutamas Jayuudiskul for
sharing her opinions and experience on Thai cooking.

Heartfelt gratitude to the descendants of Thanpuying Kleeb
Mahidhorn, i.e., Dr. Kamontip Snidvongs na Ayudhya, Dr. Kraipun
Yunibandhu, and Khun Rosalind Yunibandhu, for sharing the culinary
legacy of Thanpuying Kleeb through the *Petals of the Champaka* cookbook. In collaboration, a special admiration for Chef Tam Chudaree
Debhakam of Baan Tepa Culinary Space for creating the delightful
dishes cooked to the original taste approved by Thanpuying Kleeb family. The author was mesmerized by the total experience of the whole meal.

References

Benjakul, Soottawat, Wonnop Visessanguan, Kongkarn Kijroongrojana, and
    Pisal Sriket. 2008. "Effect of heating on physical properties and micro-
    structure of black tiger shrimp (*Penaeus monodon*) and white shrimp
    (*Penaeus vannamei*) meats." *International Journal of Food Science &
    Technology* 43 (6):1066–1072.
Bongsanid, Yaovabha. 1935. *Tumrub Sai Yaowapa* (ตำรับสายเยาวภา). Bangkok,
    Thailand: Saipunya Sakakom.

Embuscado, Milda E. 2015. "Spices and herbs: Natural sources of antioxidants – A mini review." *Journal of Functional Foods* 18:811–819. doi: 10.1016/j.jff.2015.03.005

Erickson, M. C., M. A. Bulgarelli, A. V. A. Resurreccion, R. A. Vendetti, and K. A. Gates. 2007. "Sensory differentiation of shrimp using a trained descriptive analysis panel." *LWT - Food Science and Technology* 40 (10):1774–1783. doi: 10.1016/j.lwt.2006.12.007

Goldberg, Lina. 2017. "10 of the world's best fresh markets." CNN. https://edition.cnn.com/2012/07/17/travel/worlds-best-fresh-markets/index.html.

Kethom, Wassana, Pumipat Tongyoo, and Orarat Mongkolporn. 2019. "Genetic diversity and capsaicinoids content association of Thai chili landraces analyzed by whole genome sequencing-based SNPs." *Scientia Horticulturae* 249:401–406.

Khanthapok, Patipanee, and Suchada Sukrong. 2019. "Anti-aging and health benefits from Thai food: Protective effects of bioactive compounds on the free radical theory of aging." *Journal of Food Health and Bioenvironmental Science* 12 (1):88–117.

Kongpan, Srisamorn. 2018. *Intangible Cultural Heritage Foods of Thailand* (อาหาร ขึ้นทะเบียน มรดกทางภูมิปัญญาของชาติ) Bangkok, Thailand: S.S.S.S. (บริษัท ส.ส.ส.ส. จำกัด).

Kuroda, Motonaka, and Naohiro Miyamura. 2019. "Effect of a kokumi peptide, γ-glutamyl-valyl-glycine, on the sensory characteristics of foods." In *Koku in Food Science and Physiology: Recent Research on a Key Concept in Palatability*, edited by Toshihide Nishimura and Motonaka Kuroda, 85–133. Singapore: Springer Singapore.

Lee, Chang Yong, and John M deMan. 2018. "Enzymes." In *Principles of Food Chemistry*, 397–433. Switzerland: Springer.

Mahidhorn, Thanphuying Kleeb. 1949. "Cookbook for Offsprings (หนังสือ กับข้าวสอนลูกหลาน)." In: Vajirayana Digital Library https://vajirayana.org.

Manrique, Guillermo D., and Franco M. Lajolo. 2004. "Cell-wall polysaccharide modifications during postharvest ripening of papaya fruit (*Carica papaya*)." *Postharvest Biology and Technology* 33 (1):11–26. doi: 10.1016/j.postharvbio.2004.01.007

Mizuta, S., Y. Yamada, T. Miyagi, and R. Yoshinaka. 1999. "Histological changes in collagen related to textural development of prawn meat during heat processing." *Journal of Food Science* 64 (6):991–995.

Ninrat, Nuang 1996. Chiwit nai wang (ชีวิตในวัง), Thai collection history of Southeast Asia. Bangkok: Sisara.

Phassakorawong, Thanphuying Plian. 1910. In *Tamra maekhrua hua pa*, edited by Plain Phassakorawong. Bangkok: Tonchabap. https://vajirayana.org.

Phewpan, Apiniharn, Preecha Phuwaprisirisan, Hajime Takahashi, Chihiro Ohshima, Panita Ngamchuachit, Punnida Techaruvichit, Sebastian Dirndorfer, Corinna Dawid, Thomas Hofmann, and Suwimon Keeratipibul. 2019. "Investigation of kokumi substances and bacteria in Thai fermented freshwater fish (Pla-ra)." *Journal of Agricultural and Food Chemistry* 68 (38):10345–10351.

Pisanwanich, Att. 2019. "Chinese Garlic Flooded the World, Would Thai Garlic Survive? (กระเทียมจีนท่วมโลก กระเทียมไทยจะรอดหรือไม่?)." *Thansettakij*, 15 December 2019. https://www.thansettakij.com/content/416830.

Rachanuprapan, Mom Somcheen 1890. *Tum-ra-gub-khao (ตำรากับเข้า)*. Bangkok, Thailand: Watcharin.

Schweiggert, Ralf M., Rachel E. Kopec, Maria G. Villalobos-Gutierrez, Josef Högel, Silvia Quesada, Patricia Esquivel, Steven J. Schwartz, and Reinhold Carle. 2014. "Carotenoids are more bioavailable from papaya than from tomato and carrot in humans: A randomised cross-over study." *British Journal of Nutrition* 111 (3):490–498. doi: 10.1017/S0007114513002596

Shiga, Tania Misuzu, Joao Paulo Fabi, João Roberto Oliveira do Nascimento, Carmen Lúcia de Oliveira Petkowicz, Lucia Cristina Vriesmann, Franco Maria Lajolo, and Beatriz Rosana Cordenunsi. 2009. "Changes in cell wall composition associated to the softening of ripening papaya: Evidence of extensive solubilization of large molecular mass galactouronides." *Journal of Agricultural and Food Chemistry* 57 (15):7064–7071.

Sriket, Chodsana, Soottawat Benjakul, and Wonnop Visessanguan. 2011. "Characterisation of proteolytic enzymes from muscle and hepatopancreas of fresh water prawn (Macrobrachium rosenbergii)." *Journal of the Science of Food and Agriculture* 91 (1):52–59. doi: 10.1002/jsfa.4145

Suwankanit, Chutipapha, Sawanya Pandolsook, Varaporn Vittayaporn, Kriserm Tohtubtiang, and Naruemon Nantaragsa. 2015. "Influences of different soup stocks on chemical and organoleptic properties of Tom Yum." *Journal of Science and Technology, Ubon Ratchathani University* 17 (3):56–70.

Tapsell, Linda C., Ian Hemphill, Lynne Cobiac, David R. Sullivan, Michael Fenech, Craig S. Patch, Steven Roodenrys, Jennifer B. Keogh, Peter M. Clifton, and Peter G. Williams. 2006. "Health benefits of herbs and spices: The past, the present, the future." *The Medical Journal of Australia* 185 (S4): S1–S24.

Vaclavik, Vickie A., and Elizabeth W. Christian. 2014. "Meat, poultry, fish, and dry beans." In *Essentials of Food Science*, 133–172. New York, NY: Springer.

Xoomsai, Terb. 1982. *Rattanakosin Dishes 1982*. Bangkok, Thailand: Saipunyasamakhom.

8

FINAL PERSPECTIVES

HOLGER Y. TOSCHKA AND
VALEERATANA K. SINSAWASDI

Contents

8.1 From the Perspective of the Non-Thai Editor, Holger Y. Toschka

8.1.1 Thai People and the Food Chain

The world population has increased, notably with an increasing middle class. For Thailand, the World Bank declared in 2011 that it is a "higher, middle-income country" (Xanthos 2015). The role of food for middle-income families is not just to satisfy hunger and provide nutrients but has also become a focal point. In 2018, almost 70% of all millennials shared a picture of their food online. Google showed that 13 billion people were searching for food, whereas Instagram estimated more than 250 million people per month (Taher 2019). But there is also a challenge; according to a recent estimation by the WHO, up to 1.9 billion people are overweight or obese, while 462 million people are still undernourished and lacking essential nutrients (World Health Organization 2021). Thailand was also reflecting this trend.

In 2020, during the Covid-19 pandemic, the World Bank published a GDP decline of 6.1%, the largest GDP contraction observed since the economic crisis in the late 1990s (The Nation 2019). Recently,

DOI: 10.1201/9781003182924-11

Food Industry Asia (FIA) asked Oxford Economics to evaluate the relevance of the agricultural sector to the nation. They estimated that about one-quarter of the domestic income comes from agriculture, which provides more than 40% of jobs (Oxford Economics 2021). However, though significant for the nation's economy, the agriculture sector has created some environmental concerns. For example, between 25% and 36% of carbon dioxide is emitted from the food chain (Mbow et al. 2019). The problem was undeniable when farmers burned their fields after harvest, and the air became nearly unbreathable with the massive fine particulate matter (PM 2.5) in the air.

Interestingly, despite the economic impact, Thai people are aware of the environmental issue caused by agriculture and food production. In 2021, Food Innopolis (https://foodinnopolis.or.th), the global food innovation hub focusing on research, development, and innovation for the food industry in Thailand, organized its 4th international conference on food innovation under the theme Driving a Sustainable Future Through Bio-Circular-Green Economy. The discussion showed that the agronomical research focus has shifted from maximizing crop yield to sustainable and regenerative farming practices, and also their link to dietary behavior (Churak et al. 2021).

8.1.2 Food and Health

The "invisible credentials" of good food or the importance of consumers' trust in food safety was highlighted during the Covid-19 pandemic. For most people, eating has become a mindful choice. Thai people have started to ask for more transparency and urge action against its impact. The Trade Policy and Strategy Office announced that at the end of 2019, blockchain technology would improve organic rice's traceability (The Nation 2019).

Fast-food consumption was growing, driven by millennials. Chicken, burgers, and Asian fast food chains grow initially at around 10%. But growth was slowing down mid-century, with increasing health concerns associated with fast food (Euromonitor 2019a,b). Thai consumers have become more rational and choose to buy more healthy products to cook at home, with expectations such as the immune system strengthening. During the lockdown caused by the Covid-19 pandemic, as much as 90% of consumers stayed home, and 85%

had more health concerns. In addition, almost two-thirds considered nutritional values when buying food (Kessuvan and Thongpech 2020). In a poll conducted at the end of April 2020 by Suan Dusit Rajabhat University, more than 75% of people interviewed said that they were cooking more at home. Consumers tend to have more awareness of ingredients capable of strengthening natural defenses, boosting the immune system, and supporting physical and mental resilience. The awareness is in line with the global food megatrends for global food, such as: "better for me and better for the planet," "personalized wellness," "anytime and anywhere," and "delicious experiences" (Szeto and Lingala 2020). Among Thai people, food is regarded as a primary resource to achieve these goals. A study by Nielson pointed out that Thai consumers prefer healthy eating habits to physical exercise to have a healthy lifestyle (Sachamuneewongse 2021).

Regarding meat consumption, the demand for meat and meat-based products in Asia increased with the rising proportion of middle-class households. Asian people have eaten 63% more meat in the last 20 years (from 94 to 141 metric tons). However, this trend is not as evident in Thailand. Recent statistics published by United Nations Food and Agriculture Organization show that meat consumption in Thailand is relatively low (Ritchie and Roser 2017; Food Industry Asia 2021). The average meat consumed per capita per year was only 22 kg, compared to 124 kg in the United States. What Thailand adds to its national plate is an additional 29 kg per head and year from various seafood, well above the global average of 19.9 kg. The low consumption of meat may reflect that classical Thai food is naturally meat-free, or meat is easily replaceable with vegetables or plant-based products like tofu, young jackfruit, and mushroom. Also, typical Thai meals are traditionally rich in vegetables, fruits, whole grains, herbs, and spices, quantitatively as well as qualitatively. Moreover, most meals offer high nutrients and beneficial non-nutrient functional components such as phytochemicals while low in fat and calories. Thus, the Thai kitchen has the ability to serve "richer tables" without adding meat to them.

Within the food chain, environmental and food safety issues seem to be increasing concerns for Thai people. But on the other hand, classical Thai meals, which are naturally full of vegetables, herbs, and spices, while the meat is optional, seem to fit well with current dietary recommendations. Moreover, Thai meals potentially bring not just satiety but also pleasure, as discussed throughout this book. Hence, the Thai way of cooking and eating has shown a true potential for Thai cuisine to be one of the most admired kitchens in the world, now and in the years to come.

8.2 Epilogue from the Thai Editor, Valeeratana K. Sinsawasdi

Writing this book is like I am constantly going down memory lane. I was not really on the traditional side of food preferences as a child. I almost always ate Western-style breakfast. However, mandated school lunch was a meal set that was always Thai foods and desserts during school days. Then, dinner was also always home-cooked traditional Thai dishes.

I was influenced so much by the Rajini School I attended for 12 years. Cooking was my favorite class. The sophisticated food decoration art class, e.g., fruit carving, was not very fun. However, it gave me an appreciation for the complexity, thoughtfulness, and multidimensions of Thai foods. Thanks to the school founder, Queen Saovabha Phongsri, whose vision was to educate Thai girls about traditional values (Kasempolkoon 2017). All the experiences participating in many school activities gave me a foundation to understand and appreciate Thai food and its history.

8.2.1 Inspiration from the Influencer

I wouldn't be inspired enough to write a book on Thai food if I had no experience helping my "khun-ya" (grand-aunt) around the kitchen. *Ms. Pa-ob Dhanasevi* was very proud to be born in the reign of King Rama V though she was just a few months old when the King died. Ms. Pa-ob would use old-fashioned tools like granite stone mill, granite mortar and pestle, and a little stool attached with a half-round zigzag metal blade to shred the thick hard coconut meat.

All the dishes were prepared from scratch, including coconut-based curry, and thus labor-intensive and time-consuming. Ms. Pa-ob's daily life was dealing with food, pretty much all day. She would not be willing to change the way she cooked, no matter what technology was there to offer her. I remembered conversations about the availability of processed coconut milk (canned and pasteurized formats), the packaged curry paste, or even the home delivery of the whole meal. Ms. Pa-ob always found some undesirable outcomes with the industrialized products.

The most frequent complaint was that the processed coconut milk was not fragrant enough, and it wouldn't separate into oily and watery layers as the freshly squeezed coconut milk would. This lack of emulsion break would affect the curry flavor and consistency and couldn't be corrected or adjusted at any later cooking steps. She tried to heat and stir the curry paste with oil, but the aroma obtained did not meet her expectations. A similar scenario went with the curry paste, which she could never find one that was good enough.

The attempt to use a packaged curry paste would end up with Ms. Pa-ob adding many herbs and spices till she satisfied with the sensory properties. And that modification required a lot of work anyway. I wish I had paid more attention to what she said and done. Instead, I usually get annoyed by her being so specific about every detail. The freshness of every ingredient, the maturity of herbs, the color, the aroma, etc. were all significant.

I believed Thai food was nutritious and well balanced. But at the time, I also thought that everything could be simplified (or at least minimally processed) to achieve the same outcomes. Surprisingly, writing this book made me realize just the opposite because many science principles discussed in this book were like I was defending Ms. Pa-ob's theories of Thai cooking!

8.2.2 On Health Benefits of Thai Food

In childhood, I did not particularly value Thai food for its palatability. But since then, I have already believed in the health benefits of Thai cuisine. Both of my biological grandmothers were good cooks, but I did not get a chance to spend much time with them. My paternal grandmother had rice paddies, and she also practiced traditional Thai herbal medicine as the knowledge was inherited from her family. For my master's degree at the University of Hawaii, I used Oxygen Radical Absorbance Capacity (ORAC) assay to measure total antioxidant activities curry powder and ginger drink. In the 1990s, this assay was not yet popular. The committee asked during the thesis defense why I chose to work on spices. My answer was the potential health benefits, and the concentration of bioactive compounds in spices was very high. They can be added to any food to enhance flavor and adding health benefits without replacing significant food portions. Later, at the University of Florida, I studied another plant material used as a food colorant. Sappanwood gives not only distinctive red color but also has many proven benefits.

This current book has no devoted chapter about Thai food's medicinal or health benefits. The main reason is, there are so much of interesting details and evidence that we think the topic deserves more attention, perhaps in another book devoted to Thai cuisine's health and medicinal benefits.

8.2.3 Authenticity of Thai Food

National cuisine, like the Thai one, is hard to define. When I was in the United States, new friends seemed to love mentioning their Thai food experience. One friend told me if she had to choose just one cuisine to eat for the rest of her life, she would select Thai food.

My housemate amazed me when she came back from a vacation to Thailand. She brought everything with her, like a large steamer and a full-size wok, along with all ingredients, including a big bottle of Thai fish sauce! She took a Thai cooking class during the trip and loved Thai food so much that she had to make sure she had everything needed to cook Thai food in the United States.

But I also found that Thai food means different things to each person. Many friends brought me to several "very authentic" local Thai restaurants. But, to me, there were only a few resemblance elements, so I could barely name those dishes. In other words, the food was not that good. However, just because of spiciness, my friends seemed to think the foods were delicious and "authentic."

During my coursework at the University of Florida, I listened to Dr. Linda Bartoshuk's lecture on supertasters and how each person genetically has a different sensitivity and perception of tastes and smells. Participating in her study, I found that I am a supertaster. This status did not come with a surprise as it was just like what she had explained in her research for most other Asians. With observations, I agreed with Dr. Bartoshuk's theory that Asian people seem to enjoy the food and be satisfied at a lower caloric content. So, it is like I am living in a more colorful and vivid food world.

Then later when I studied what Dr. Charles Spence (2017) theorized about how a surprise can be either pleasant or disgusting, the theory helped me to puzzle it out. Thai cuisine caught a lot of people by surprise at the first try because of the unusual combination of herbs, spices, and burning sensation. That surprise is mostly considered pleasant. They like the food even though they may not get the same intensity and varieties of the tastes and smells the way Thai people perceive. So, I have concrete expectations for what a Thai dish should taste, and I will like the food only if it meets my expectations. However, my foreign friends may perceive different flavors, but they enjoy Thai meals anyway.

8.2.4 Food as Soft Power

I have been familiar with Thai food as available nowadays since childhood. Thus, when I researched Thai food history, I was amazed by how significant elements of Thai food culture were invented, mostly just in the last decade. In 1861–1862, Graf Friedrich zu Eulenburg (Count Friedrich Albrecht zu Eulenburg) a Prussian (German) diplomat, visited Siam (Thailand). Count Eulenburg met King Rama IV, and the crown prince (future King Rama V), so it was a considerable transition period before the modernization started (zu Eulenburg 1862).

There was no admiration or even anything positive about Thai cuisine in the record. Back then, Thai people had a betel chewing tradition, which caused chewers' teeth to turn black and look "dirty." The habit required spitting throughout the day, so elite households that the diplomat visited openly showed the potty in every corner (the tradition was so highly valued that the betel kit and potty were status symbols). Eggs at breakfast were from alligators or turtles, and fruits did not taste good. People would eat on the floor with bare hands, including serving high-ranking guests. What Count Eulenburg described was no different from other foreigners who visited Siam before him, e.g., Simon de la loubère, a French Diplomat who arrived in 1687–1688 (Sujachaya 2017). Thai people ate rice, fish, fermented fish, chili paste, and plenty of assorted vegetables and fruits.

Count Eulenburg mentioned that the Thai women's hairstyles and clothing were almost identical to men's. However, fashion was about to change dramatically right after his visit as one of several attempts Siam had endured to appear more civilized to the West. The reason

Siamese women dressed like men was because we had several wars. Thai women also served in the army as warriors. There are monuments to honor Thai women's bravery, such as *Thao Thep Krasattri*, *Thao Si Sunthon*, and *Thao Suranari*. The unisex outfit was also a Siamese lady strategy to make it easier to disguise as a man.

Then, during modernization in King Rama V's reign, ladies still wore the *chong-kraben* (one piece of cloth wrapped around the waist, tied, rolled, and folded to wear like pants) and short hair but adopted Victorian-style lace shirt, jewelry, stocking, and low heel shoes. The *Sabai* garment, which had been used to wrap around the upper body, was changed to a diagonal drape from a shoulder and tied a knot at the hip. The fashion of this period was short-lived, but it served as a perfect example of how Thai people ingeniously embraced and blended elements from other cultures. Food culture also clearly showed this direction, i.e., the fusion of heritage and foreign elements.

Changes were well received, and in 1897, Siamese King and Queen visited several countries in Europe, including Germany, as Royal guests. The several strategies to modernize the country and people worked out successfully. Siam has a legacy of maintaining independence throughout the colonization period. The *Mae-Krua-Hua-Pa* cookbook of Thanphuying Plian Phassakorawong (1908) mentioned in many chapters of this book is a vivid evidence of how food culture was modernized and fused with other food cultures. Thanphuying Plian discussed not only food recipes but also science, technology, nutrition, sanitation, and even supply chain management.

Studying the *Mae-Krua-Hua-Pa* book was an eye-opener for me on how rapid, efficient, innovative, and ingenious Thai people were. Hence, I expanded my interest into neighboring countries and Europe during the late Victorian era. Thailand rapidly changed during this period, with electricity, cars, trains, and abolition of slavery.

Thai people turned around the food culture within one generation, from personal hygiene to cutlery to table settings. Then shortly, *tom-yum-koong*, pad thai, and varieties of foods were founded. Thai cuisine dynamics may be regarded as a strategy, a soft power, a gastrodiplomacy, or purely a national intangible heritage—nonetheless, my sincere gratitude and admiration for our ancestors' hard work and dedication.

More recently, the Covid-19 pandemic, the industrialization, and technology disruption have induced changes in our food markets, kitchens, cooking, and dining. Therefore, I think documentation on the food industry and food culture transition is highly motivating and may be of interest to the general public in the future.

References

Chavasit, V; Kasemsup, V; Tontisirin, K. 2013. "Thailand conquered under-nutrition very successfully but has not slowed obesity." *Obesity Reviews* 14:96–105.

Churak, Piyanit; Sranacharoenpong, Kitti; Mungcharoen, Thumrongrut. 2021. "Environmental consequences related to nutritional status of Thai populations." *Journal of Public Health* 29 (4):879–884.

Euromonitor. 2019a. Fast Food in Asia Pacific https://www.euromonitor.com/article/fast-food-asia-pacific

Euromonitor. 2019b. Top Consumer Trends Impacting Health and Nutrition (Part 2) https://go.euromonitor.com/whitepaper-health-and-nutrition-2019-health-and-nutrition-survey.html

Food Industry Asia. 2021. The Voice of the Asian Food Industry | Food Industry Asia. https://foodindustry.asia/hubfs/FIA-Oxford%20Economics%20-%20Climate%20Change%20and%20Food%20Prices%20in%20Southeast%20Asia.pdf?hsLang=en

Interactive Schools. 2018. "50 million users: How long does it take tech to reach this milestone?—Interactive Schools." *Interactive Schools*, 8 February. https://blog.interactiveschools.com/blog/50-million-users-how-long-does-it-take-tech-to-reach-this-milestone

Kasempolkoon, Aphilak. 2017. "When "Royal Recipes" became more widely known: Origin and development of "Royal Cookbooks" in King Rama V's reign to girls' schools' establishment." *Vannavidas* 17:353–385.

Kessuvan, Ajchara; Thongpech, Arisara. 2020. "Towards the new normal life-style of food consumption in Thailand." *FFTC Agricultural Policy Platform.*

Mbow, Cheikh; Rosenzweig, C, Barioni, LG; Benton, TG; Herrero, M; Krishnapillai, M; Liwenga, E; Pradhan, P; Rivera-Ferre, MG; Sapkota, T. 2019. "Food security." In: Climate Change and Land: An IPCC Special Report on Climate Change, Desertification, Land Degradation, Sustainable Land Management, Food Security, and Greenhouse Gas Fluxes in Terrestrial Ecosystems. *Geneva: Intergovernmental Panel on Climate Change (IPCC).* https://www.ipcc.ch/site/assets/uploads/sites/4/2019/11/08_Chapter-5.pdf

Oxford Economics. 2020. "Mapping Asia's food trade and the impact of COVID-19." *Food Industry Asia.* https://www.researchgate.net/publication/349363329_Food_system_resilience_and_COVID-19_-_Lessons_from_the_Asian_experience/link/60444e4b299bf1e0785f71a6/download

Oxford Economics. 2021. The Economic Impact of the Agri-Food Sector in Southeast Asia. https://foodindustry.asia/hubfs/FIA-Oxford%20 Economics%20-%20Climate%20Change%20and%20Food%20 Prices%20in%20Southeast%20Asia.pdf?hsLang=en

Phassakorawong, "Thanphuying Plian." 1908. In: *Tamra Maekrua Huapa (ตำราแม่ครัวหัวปาก์)*, edited by Plian Phassakorawong. Bangkok: Vajirayana Digital Library. https://vajirayana.org.

Reddy, K Srinath. 2016. "Global burden of disease study 2015 provides GPS for global health 2030." *The Lancet* 388 (10053):1448–1449.

Ritchie, Hannah; Roser, Max. 2017. "Meat and dairy production." *Our World in Data.*

Sachamuneewongse, Siriporn. 2021. Thais Opt for Health Food Over Exercise. https://www.bangkokpost.com/business/1658872/thais-opt-for-health-food-over-exercise

Spence, Charles. 2017. *Gastrophysics: The New Science of Eating*: UK: Penguin.

Sujachaya, Sukanya. 2017. "Thai cuisine during the Ayutthaya period." *Humanities Journal* 24 (2): 1–29. https://www.tci-thaijo.org/index.php/abc/article/view/106404

Szeto, Doreen; Lingala, Anu. 2020. "Digesting change: A global look at consumer trends in food and beverage." *Kantar.*

Taher, Alina. 2019. Instagram: Key Global Statistics 2019. https://blog.digimind.com/en/trends/instagram-key-global-figures-2019

The Nation. 2019. Blockchain Technology to be Used in Traceability of Organic Rice. https://www.nationthailand.com/in-focus/30378722

Willett, Walter; Rockström, Johan; Loken, Brent; Springmann, Marco; Lang, Tim; Vermeulen, Sonja; Garnett, Tara; Tilman, David; DeClerck, Fabrice; Wood, Amanda. 2019. "Food in the anthropocene: The EAT–lancet commission on healthy diets from sustainable food systems." *The Lancet* 393 (10170):447–492.

World Health Organization. 2021. "Malnutrition." *World Health Organization.* https://www.who.int/news-room/fact-sheets/detail/malnutrition

Xanthos, Dimitris. 2015. "Country in focus: Economic transition and non-communicable diseases in Thailand." *The Lancet Diabetes & Endocrinology* 3 (9):684–685.

zu Eulenburg, Graf Friedrich. *Siam in briefen von Graf Friedrich zu Eulenburg 1862*. Translated by Otrakul, Ampha. Bangkok, Thailand: Chulalongkorn University Press, 2021.

Appendix A
First Memory of Thai Cuisine

NATE-TRA DHEVABANCHACHAI AND VALEERATANA K. SINSAWASDI

A.1 Background

Foreigners who had lived in Thailand were asked to think about the day they tried Thai food for the first time (regardless of the place or the dish). Then they were asked the simple open-ended question *what can you recall about that day?*.

The participants of this survey were expatriates. This demographic of people were chosen as participants because they were foreigners with unlimited chances to try Thai food while residing in Thailand. The short open-ended question allowed the participants to express their recollection freely, how vivid the memory is retained, and possibly, their current attitude and habit after some time has passed.

Fourteen foreigners responded to this question, and their unabridged responses via text communication are listed in the details section.

A.2 Initial Findings

Responses were in line with those comments and reviews commonly found on any food reviews and travel website. The first try of the Thai dish brought surprises because the combination of tastes and aroma were new to the person. Though perceived flavors were beyond expectation, they considered the experience to be pleasant and memorable. Even for someone already familiar with spices, Thai food still gave pleasure because of the surprisingly dry spices with fresh herbs. Although the perceived flavors were intense and somewhat irritating (hot, burning), like riding a roller coaster, the whole meal was harmonious. The uniqueness of Thai food has become a signature experience motivating tourists to visit Thailand. Overall, the pleasant surprise of the first try also created a long-lasting memory and easily tempted a person to acquire more.

Not everyone likes Thai food from the first try, but it is possible to turn more affection with repeated consumption. However, if the exposure occurred at a very young age, the surprising experience with new flavors tends to be disgust rather than pleasure. Young Thai children also require some time and exposure to develop their palate for spicy dishes, but no rigid ritual or rushing. That is why *sum-rub* with an assortment of dishes is very important as it gives freedom for kids to explore. Interestingly, with more exposure, a person can learn to like a more complex and intense dish such as the shrimp-paste chili dip (*nam-prig-ga-pi*, น้ำพริกกะปิ). Nonetheless, even without surprising herbs and spices, unique Thai food elements, such as the crunchiness of fresh bean sprout, can bring fond memories associated with Thai meal pleasure.

A.3 Details of Responses

A.3.1 Responder 1: An explosion of new flavors

The 1st day I ate Thai food was when I started my internship at Banyan Tree Phuket in Aug 1998. I was so excited to be in Thailand and loved all the different smells, sights and sounds. I got to stay at one of the villas and had never been so impressed before. I think my first meal was at the hotel canteen and I remember it as an explosion of new flavors in my mouth. Coming from Europe especially the spices left a memory hahaha. The colors of the food were vibrant and the flavors unlike anything I had ever eaten before.

A.3.2 Responder 2: A roller coaster of flavors

My first exposure to Thai flavors was in the US. My girlfriend (who eventually became my wife), took me to a Thai restaurant. I remember the scents of my first Thai Tea, the unusual combination of ingredients of my Beef Salad and the power of my first hammock. It was unlike anything I tried before. Dishes didn't arrive in a particular sequence, no protocol! No fuss. The fragrant herbs, curries and spices, the association of sweet & sour, chili and salt was like a roller coaster I had never experienced before: Flavors did not "mix" they built layers across the palate, leaving a spicy hot lingering finish. The contrast of textures, the light crunch of pounded roasted rice, the scents of mint and coriander leaves, the juicy ripe mango on a bed of warm shinny sticky rice, the crunch of roasted seeds, the creamy coconut sauce… every bite was exoticism at its paroxysm!

A.3.3 Responder 3: Variety of flavors

My family all love Thai food, I remembered the 1st time was eating out with friends and at that time I haven't been to thailand yet. My impression is that it is such a delicious foreign food which has mix with varieties flavors: sour, spicy, deep fried, lemon grass in the soup or as seasoning in the dishes, to create special taste.

A.3.4 Responder 4: Stronger, spicier, sweeter

My first time trying Thai food was an unforgettable experience because I am not a big fan of spicy foods and every time I crave something spicy, I think of Thai's foods. Thai and Cambodian's foods have very similar taste, but Thai's foods taste stronger, sweeter, and spicier. I love Thai foods a lot, I love almost every dishes. And every time I think of Thai foods, I always recall my great memories with Thai people too.

A.3.5 Responder 5: Attracted by the spicy dish

The first thai food impresses me well is Tom Yum Gong soup, which i am still into now. I still remembered that the first time i tried it was in a small restaruant nearby the sea in Samui Island. Also that was the first time i experienced the real thai food. I went there with my classmates, the weather that day was great, but we couldnt help to sweat, cuz the soup was really spicy for us at that time. And after that, I was literately attracted by it. Until now, i often buy the Tom Yum ingredients online and cook it myself, but the flavour is not tasty as the one i tried in Thailand.

A.3.6 Responder 6: Shocking moment

My first experience of Thai food is when I ate it at the Thai restaurant in Osaka with my friend. We tried to eat sea food salad with Thai taste chili sauce that was so delicious as I love the transparent noodle. And Tom yam kun which is Thai traditional hot soup, that was my shocking moment I never had tasted like such hot soup. But I felt Thai culture very much and I loved it since then. I constantly eat Thai food at restaurants in Osaka. We have very good restaurants cooked by Thai chef.

A.3.7 Responder 7: With herbs, thai cuisine is different from dry spices of Indian cuisine

I first came across Thai food in Sydney, Australia and my first impression was one of interest and intrigue - I ordered Yum Pla Duk Foo, Green Curry, Stir-fried Kale and Jasmine Rice - Overall taste was sweet yet spicy - delicious but, an interesting combination. I was particularly interested because, as a professional Chef trained in French cuisine, which is rich and creamy; sweet and savory are two separate flavors and rarely mixed. However, when I started work in Thailand, I realized that the Australian version of Thai cuisine was "anglicized" to the western palette i.e. spice reduced with sweetness to make it more attractive. Having said that, in later trips, I discovered some great, authentic Thai restaurants in Sydney which were more reflective of the real "Thai cuisine".

In summary, my 18 years in Thailand has since taught me that the freshness of the herbs and spices are very different than the dry spices of South Asian cuisines like India (my native cuisine), Sri Lanka etc. The light cooking techniques of stir-fry's where vegetables are still crunchy and full of natural nutrients is light compared to some cuisines where the vegetables are overcooked. It is just one of the most simple yet intricate cuisines; Fresh yet technical; small spoonfuls pack a big impact. We eat more Thai food at home than any other cuisine.

A.3.8 Responder 8: Tourist attraction

I would like to share the first day I tried Thai food, it's Pad Thai, I was amaze by the garnishments which are more than ten. It is an iconic meal which every tourist tried when they arrive in Thailand. It's kind of mission to complete in their bucket list. My impression on thai cuisine was details oriented, pay attention to cost-efficient ingredients, and taste great.

A.3.9 Responder 9: Unforgettable "Taste of Siam"

Thai Food was a culinary art created with passion and historical background, a cuisine in which food was curated, prepare and made with authenticity that demonstrate the elements used to create the food. It displayed the quality, the aromatic ingredients by using medicinal herbs and spices that makes it unique and flavorful, the food appearance that presented make your palate to indulge in with a kick of spiciness & sweetness that make an harmony to end.

It was 15 years ago when I introduced to Thai Cuisine at Singapore. It was on my Internship "International Practicum Training Program" in which I selected to be part of. Back then I was assigned to be an intern at the two establishments of the group of companies in which one is a Thai Restaurant named as Lerk Thai® "The Taste of Siam", a famous restaurant that offers authentic Thai cuisine. I will not mention only one food but rather Thai food that I tried on first time I step in as an intern, to mention and name those food are the following such as Tom Yum, Nua Phad Prik Thai Dum (Beef with Black Pepper Sauce), Pla Krapong Tod Nam Pla (Deep Fried Seabass with Mango Salad), Gaeng Massaman Nua (Massaman Beef Curry), Pad

Kra Prao(Minced Meat with Thai Basil and Chilli), Plaa Raad Prik
(Deep Fried Fish "Grouper Fish" with Sweet Chili Sauce), for the
sweets of course the Khao Niew Mamuang (Mango with Sticky
Rice and Coconut Milk) and some Herbal beverages created to offer
to our guests. Most of the food I mention was the food that I taste
mostly on my first taste of Thai food especially the Tom yum which
like a soup same as "Sinigang" a tamarind soup based we have in
Philippines, Tom yum this hot soup when you sip that sour spicy
sweet makes the warmth feeling especially in cold rainy days. The
Deep fried Sea bass with Mango Salad this food has a good combination of earth and sea ingredients makes the complement of to
both fresh fish and juicy mango. To share with you as the Phad Thai
and Beef Dish impact my craving so I've tried to cook the Beef with
Black Pepper Sauce (Nua Phad Prik Thai Dum) just by recalling all
ingredients and find in the market when I back to my hometown in
Philippines, my family like and loved it much especially the Phad
Thai that I make and create my own version so they requested to
cooked it on special occasion as new dish on the table and as it was
new to them glad it was approved, appreciate the taste though it is
not as Authentic as what I've taste in Singapore that time but almost
near (hahahaha).

Most of that Thai food I've tried at Singapore was really "great taste
& awesome" I may not be a known Chef but as a person who love to
cook 😑☺, eat and appreciate food and the history and tradition behind
it makes you feel good.

Everytime you taste it you'll recognize the finest selected ingredients they used to produce that dish and offer to serve with full of
details. Then I just remember during my period as an intern, my boss
asked me would you like to visit Thailand with me of course I was
amazed as I wanted to "the first thing come to my younger mind was
to see the live elephants" 😄😄☺☺ but then culture, tradition and of
course the Thai Food was one of the reason my boss wanted to bring
me and visit Thailand 15 years ago for me to have more experience,
skills and extra knowledge about Thailand, unfortunately it not hap-
pens that time ☺.

All in all that first time that that I taste the "Taste of Siam" was
unforgettable and will remain as an open eye to me by that age that
I travelled and Asian Thai Cuisine was full of history and perfectly

made by talented chef with great skills that proved to be offer to all nations. Ohh it makes me reminiscing the first time I was landed to Singapore, taste and experience not only Singaporean, Malaysian, Chinese mostly Thai cuisine.

A.3.10 Responder 10: Learn to like after many tries

Tom Yum Kung is my first Thai food in my life. I remember it tasted very unique and unfamiliar, and I didn't like it. It had a strong sour and spicy taste, and I didn't prefer since any Korean food do not taste like this. But after living in Thailand for several years, I got used to the taste of Tom Yum Kung and started to think it was delicious, so I remember buying it by myself often.

A.3.11 Responder 11: As a young child, not really like Thai dish

The first time i had thai food as far as i could recall would be was when i was a young child, tbh not the best experience since my taste buds were custom to indian food and not the slightly more spicier and more richful flavours of my first thai dish (phak boong with rice)

Since my options are limited beinf a vegetarian, i felt personally after havin a lot of phak boong and phak phak my taste buds developed its rich in nutrients and tasty flavours of the soya sauce and other thai spices

Till date i would say its a neck to neck competition for me between thai food and indian food although both of them given at any given day is a tasty meal with its enriched flavours and smell the food offers. If asking about my favorite food right now. It would be kaeng kiew wan.

A.3.12 Responder 12: First try at young age, do not like the combination

My first ever Thai food I've eaten was in Singapore(where I grew up) It was a mid-high priced range buffet style of all you can eat Thai food. It was located part of some fancy hotel. (which I do not remember)

The first impression of many of the Thai food I've seen on the table top was that it was very colorful and decorative. They would use many plant leaves to decorate their foods, in shapes of flowers and animals. I've seen many bright red, orange, green and yellow dishes which weren't very common in Singaporean food. I remember being picky as a kid with what I was going to eat as the dishes all seemed uneatable with knowing many unknown ingredients are being used.

I also remember the restaurants using pineapples in the fried rice which really surprised me. I thought to myself, why would they even try mixing carbohydrates and fruit together. I would separate the two and eat only the rice.

The deserts was something I hated in Thai food. All I wanted was something related to chocolate. Most of the green looking jelly just tasted like coconuts and pandan which was really bad for me.

There was also a mountain of fruits variety you could eat. Fruits were being carved and decorated which was extremely artistic.

The first Thai food for me was something I thought I'd never go back eating for. I believe it was some sort of culture shock for me. I was well exposed to Japanese and Chinese food which I believe rarely uses fruits in their dishes as well as coconut milk.

A.3.13 Responder 13: Intense flavour increases appetite

To be honest, I am a Thai-born Chinese and I have been living in Thailand since I was born, so I am familiar with Thai food. However, I am happy to answer in regards to your question. So, let me tell you the first time I tried Nam Prick Kra Pi. It was about years ago. I remember the flavor from krapi was extremely intense and it increased my appetite. As a result, I ate three times more than I usually do and then gained a lot of weight.

A.3.14 Responder 14: Nostalgia

My first memories of Thai food is when my family and I lived in Saudi Arabia.

Due to the nature of the middle-east, there were almost no Thai restaurants in the entire area, so the only times I experienced Thai cuisine was from my mothers cooking. Similarly, Thai ingredients were rare to come by at the time.

Pad Thai was one of the first meals I can remember eating that my mother made, and I remember the crunch of beansprouts and peanuts that my mum would put into the meal. I still have memories of helping my mother squat under our small apartment sink in Saudi Arabia and growing small little buckets of bean sprouts for our own home-made Pad Thai.

To me, Pad Thai was more than just a common Thai dish but it represented one of the only real connections that we had to home; as we grew and ate our own homemade bean sprouts during our many isolating months in Saudi Arabia.

APPENDIX B
PAD THAI RECIPE

MALLIKA TUMWATTANA AND CHOMPLOY LEERAPHANTE

B.1 Ingredients

Part 1. Pad Thai *Sauce*

1.	Sugar	1	Tablespoon (15 g)
2.	Fish sauce	1	Tablespoon (12 g)
3.	Tamarind paste	1	Tablespoon (15 g)
4.	Water	2	Tablespoon (20 g)
5.	Palm sugar	1	Tablespoon (20 g)

Part 2. Pad Thai *Ingredients*

1.	Dried rice noodle	100	Gram
2.	Shrimps	4	Pieces
3.	Palm oil	1	Tablespoon (15 g)
4.	Shallots (diced)	1	Tablespoon (5 g)
5.	Salted turnip (diced)	2	Teaspoon (5 g)
6.	Dried shrimp	1	Tablespoon (7 g)
7.	Yellow tofu (sliced, fried)	2	Tablespoon (25 g)
8.	Egg	1	Egg
9.	Lime juice	½	Teaspoon (2 g)

10.	Peanuts (crushed)	2	Teaspoon (5 g)
11.	Chili pepper	1	Teaspoon (2 g)
12.	Bean sprout	30	Gram
13.	Chinese chives	7	Gram

Part 3. Omelet

| 1. | Egg | 2 | Eggs |
| 2. | Palm oil | 1 | Tablespoon (8 g) |

Part 4. Finishing and Garnish

1.	Banana blossom	1	Stem (50 g)
2.	Bean sprout	35	Gram
3.	Chinese chives	3	Stem (8 g)
4.	Asiatic pennywort	3	Stem
5.	Lime (wedge)	1	Piece

Figure B.1   This recipe is courtesy of Ruen Mallika Restaurant, a Michelin Bib Gourmand restaurant for Thai food. (Reprinted with permission from the rightful owner, Ruen Mallika.)

B.2 Instructions

1. Prepare the garnish items on the plate. Fold the banana blossom leaf inward towards the stem, working from the outer most leaf, folding 3–4 layers. Place it on the plate, along with the bean sprout, *Asiatic pennywort*, Chinese chives, and a slice of lime.
2. Fry the shrimp in the electric fryer for 10 seconds.
3. Boil the dried rice noodle in boiling water for 20 seconds then drain it.
4. Heat the palm oil in the wok on the medium heat.
5. When the wok is hot, add the shallots, salted turnip, yellow tofu, and stir-fry them.
6. Add the egg to the wok, and stir until cooked.
7. Add the dried rice noodle and Pad Thai sauce, lime juice, shrimps, dried shrimp, peanuts, chili pepper, mix and stir fry everything.
8. Stir quickly the bean sprout and Chinese chives on the high heat.
9. Crack two eggs in a bowl and mix them well with fork.
10. Heat the wok on medium heat. When it is hot, spread the oil all around the pan and pour out the excess.
11. Add slowly the beaten egg on the wok, and spread it throughout the wok like a crepe.
12. When one side is cooked, flip over the other side. Make sure both sides of the omelet are cooked. Then set aside.
13. Spoon the cooked noodle mixture into the center of the wok and scatter over the peanuts, then fold over the sides of egg wrap into the center to cover the noodles and form a parcel. Flip the egg-wrapped Pad Thai over in the wok to briefly seal the folded ends, then slide it onto a serving plate.
14. Slit the egg wrap into four slides and open them up. Add the shrimp on the top of the plate.

Index

Note: **Bold** page numbers indicate tables in the text

9 780367 763350